Reasonable Books = Reasonable Prices

Irina Kleyman

Exceptional for Home Schooling

Math Help Grade 4

AF882381

First Edition

Reasonable Books = Reasonable Prices

Irina Kleyman

Exceptional for Home Schooling

Math Help Grade 4

I dedicate this book to my children and grandchildren.

Acknowledgments
Special thanks to Michael G LaFratta, Vladimir Kleyman,
who has assisted in bringing all 10 books to market.

Author's note: In the many years that I've taught mathematics, I have been informed by many of my students that my methods for teaching have helped them tremendously in learning math. I have written these books in order to continue to help others learn math effectively.

Math Help Grade 1
Math Help Grade 2
Math Help Grade 3
Math Help Grade 4
Math Help Grade 5
Math Help Grade 6
Math Help Grade 7
Math Help Grade 8
Math Help for College (Summary Grades 1 – 5)
Math Help for University (Basic, Algebra, Geometry)

Copyright © 0169 2020
ISBN: 978-1-7349699-3-1 Paperback

All rights reserved. No part of this book may be reproduced or transmitted in any form or by any means, electronic or mechanical, including photocopying, recording, or by any information storage and retrieval system, without permission in writing from the copyright owner. Printed in the United States of America.

Syllabus Math 1, 2, 3, 4

Grade 1 # 1 – 4   Grade 2 # 5 – 8   Grade 3 # 9-12   Grade 4 # 13 – 16

Syllabus includes topics, what students need to know.
Unknowns are numbers, which we need to find.

#	Grades	Topics. Advanced Problems (A-4), Pages (p)
	Chapters	Basic Problems (B), Chapter (C), Lesson (L)
1	1-1	One-Digit Numbers (B)
2	1-2	Two-Digit Numbers (B)
3	1-3	Three numbers (B)
4	1-4	Finding unknowns (A-1)
	Grade 1	Final Exam 1 B, C 4, L 12 # 1-20 odd, p 94
5	2-1	Addition or Subtraction Numbers (20–29) (B)
6	2-2	Addition or Subtraction Numbers (50–59) (B)
7	2-3	Addition or Subtraction Numbers (30–39) (B)
8	2-4	Finding unknowns (A-2)
	Grade 2	Final Exam 2 B, C 4, L 17, # 1 – 20 odd, p 106
9	3-1	Addition or Subtraction Numbers (60–69) (B)
10	3-2	Addition or Subtraction Numbers (90–99) (B)
11	3-3	Addition or Subtraction Numbers (70–79) (B)
12	3-4	Finding unknowns (A-3)
	Grade 3	Final Exam 3 B, C 4, L 19, # 1 – 20 odd, p 121
13	4-1	Three-Digit numbers (B)
14	4-2	Multiplication numbers (B)
15	4-3	Division numbers (B)
16	4-4	Finding unknowns (A-4)
	Grade 4	Final Exam 4 B, C 4, L 12, #1-10 all, p 133

Parents, Tutors, Students Middle School, High School, Colleges, University can help primary school students. Please read slowly math books to primary school students.

Grade 4

www.mathbooksbyirina.com

Chapter 1 Three-Digit Numbers

Grade 4 Chapter 1 Contents
Three-digit numbers
Lesson 1 – 10
Basic Problems

Work with lesson 1, read rules and examples, 2) Solve the problems from contents, 3) Check answers in the end of chapter 1, 4) Work with the next lessons 2, 3, 4, 5, …, 5) Solve Test 1 A, 6) Grading Scale: A =10 points, B=9points, C=8 points, D=7 points, F=0-6 points, 7) If you solve 10 problems correctly, you get an A, 8) If have an A, you can move on to next chapter, 9) If your grade is lower than an A, work with the mistakes, then solve Test 1B, 10) When you finish all of the chapters, solve Final Exam in last chapter.

Grade 4 Chapter 1 Lesson 1
One-digit and two-digit numbers

One-digit numbers
Count forward from 1 to 9 — 1, 2, 3, 4, 5, 6, 7, 8, 9
Count backwards from 9 to 1 — 9, 8, 7, 6, 5, 4, 3, 2, 1
Equal = Less than < Greater than >
Compare one-digit numbers (put <, or >, or =)
5 = 5 0 < 1 5 < 8 6 > 4 9 > 8
Two-digit numbers
Digits from 0 to 9 — 0, 1, 2, 3, 4, 5, 6, 7, 8, 9
One-digit numbers have one digit. This is numbers from 0 to 9
Two-digit numbers have two digits. This is numbers from 10 to 99
Digit from the left called tens. Digit from the right called ones
Comparison one-digit and two-digit numbers
One-digit numbers always less than two-digit numbers 9 < 11
Two-digit numbers always greater than one-digit numbers 19 > 8
To compare two-digit numbers:
Compare the tens, then compare the ones 12 < 18 18 > 12
Two operations with numbers. Numbers 80 – 99
Three numbers can add in any order: 11+14+19=11+19+14=30+14=44
To do two operations with numbers, work from left to right
To subtract three numbers, work from left to right: 18 – 2 – 8 = 16 – 8 = 8
To add and subtract numbers, work from left to right: 18 + 8 - 2 = 26 - 2 = 24
To subtract and add, work from left to right: 18 - 9 + 8 = 9 + 8 = 17
Add and Subtract Numbers: 68 – 29 = 68 – 20 – 9 = 48 – 9 = 39
39 + 29 =30 + 9 + 20 + 9 = 30 + 20 + 9 + 9 = 50 + 18 = 68
Properties of zeros
Two numbers can add in any order. The sum will be the same.
2 + 3 = 5 3 + 2 = 5
The sum of a number and zero is the same number:
5 + 0 = 5 0 + 5 = 5
Difference any number and zero is the same number: 5 – 0 = 5
Difference two the same number is zero: 5 – 5 = 0
To check addition use operation subtraction
2 + 3 = 5 Check 5 – 2 = 3 or 5 – 3 = 2
To check subtraction use operation addition and subtraction
5 – 2 = 3 Check 2 + 3 = 5 or 5 – 3 = 2

Grade 4 Chapter 1 Lesson 1
Addition and Subtraction Numbers
Addition

First number in addition called the addend	2
Second number in addition called the addend	3
Result of addition called the sum	5

Subtraction

First number in subtraction called the minuend	5
Second number in subtraction called the subtrahend	2
Result of subtraction called the difference	3

Add or Subtract Numbers in row

# 1	18+12=10+8+10+2=20+10=30	# 11	30–22=30-20-2=10-2=8
# 2	22 + 9	# 12	31 – 12
# 3	28 + 6	# 13	34 – 27
# 4	18 + 17	# 14	35 – 19
# 5	29 + 7	# 15	36 – 9
# 6	25 + 8	# 16	33 – 14
# 7	26 + 6	# 17	32 – 24
# 8	23 + 16	# 18	39 – 28
# 9	29 + 9	# 19	38 – 19
# 10	28 + 9	# 20	37 – 18

Add or Subtract numbers in column

# 21 23 +17	# 22 24 + 18	# 23 25 +19	#24 29 + 17	#25 29 +19
#26 12 +19	#27 26 + 17	# 28 26 + 19	# 29 28 +19	# 30 35 +14
# 31 41 -28	# 32 43 -29	# 33 45 -16	# 34 47 - 19	# 35 49 - 15
# 36 40 -15	#37 42 -26	# 38 44 - 17	#39 46 -18	# 40 48 - 29

Grade 4 Chapter 1 Lesson 2
Put plus or minus with numbers
Review 1 A

Put + or – between numbers that each problem was correct

20 ? 60 = 80   20 + 60 = 80

60 ? 20 = 40   60 - 20 = 40

	a	b	c	d
# 1	20 ? 40=60	30 ? 30=60	40 ? 20=60	50 ? 10=60
# 2	60 ? 10=50	70 ? 10=60	80 ? 10=70	90 ? 10=80
# 3	21 ? 42=63	31 ? 32=63	41 ? 22=63	51 ? 12=63
# 4	61 ? 12=49	71 ? 12=59	81 ? 12=69	91 ? 12=79
# 5	22 ? 44=66	32 ? 34=66	42 ? 24=66	52 ? 14=66
# 6	62 ? 14=48	72 ? 14=58	82 ? 14=68	92 ? 14=78
# 7	23 ? 46=69	33 ? 36=69	43 ? 26=69	53 ? 16=69
# 8	63 ? 16=47	73 ? 16=57	83 ? 16=67	93 ? 16=77
# 9	24 ? 48=72	34 ? 38=72	44 ? 28=72	54 ? 18=72
# 10	64 ? 18=46	74 ? 18=56	84 ? 18=66	94 ? 18=76

Put + or – between numbers that each problem was correct

20 ? 70 = 90   20 + 70 = 90

90 ? 20 = 70   90 - 20 = 70

	a	b	c	d
# 11	25 ? 41=66	35 ? 31=66	45 ? 21=66	55 ? 11=66
# 12	65 ? 11=54	75 ? 11=64	85 ? 11=74	95 ? 11=84
# 13	26 ? 43=69	36 ? 33=69	46 ? 23=69	56 ? 13=69
# 14	66 ? 13=53	76 ? 13=63	86 ? 13=73	96 ? 13=83
# 15	27 ? 45=72	37 ? 35=72	47 ? 25=72	57 ? 15=72
# 16	67 ? 15=52	77 ? 15=62	87 ? 15=72	97 ? 15=82
# 17	28 ? 47=75	38 ? 37=75	48 ? 27=75	58 ? 17=75
# 18	68 ? 17=51	78 ? 17=61	88 ? 17=71	98 ? 17=81
# 19	29 ? 49=78	39 ? 39=78	49 ? 29=78	59 ? 19=78
# 20	69 ? 19=50	79 ? 19=60	89 ? 19=70	99 ? 19=80

Grade 4 Chapter 1 Lesson 2

Rewrite, Read, Memorize Numbers

Numbers from 1 to 10 by 1	1, 2, 3, 4, 5, 6, 7, 8, 9, 10
Numbers from 2 to 20 by 2	2, 4, 6, 8, 10, 12, 14, 16, 18, 20
Numbers from 3 to 30 by 3	3, 6, 9, 12, 15, 18, 21, 24, 27, 30
Numbers from 4 to 40 by 4	4, 8, 12, 16, 20, 24, 28, 32, 36, 40
Numbers from 5 to 50 by 5	5, 10, 15, 20, 25, 30, 35, 40, 45, 50
Numbers from 6 to 60 by 6	6, 12, 18, 24, 30, 36, 42, 48, 54, 60
Numbers from 7 to 70 by 7	7, 14, 21, 28, 35, 42, 49, 56, 63, 70
Numbers from 8 to 80 by 8	8, 16, 24, 32, 40, 48, 56, 64, 72, 80
Numbers from 9 to 90 by 9	9, 18, 27, 36, 45, 54, 63, 72, 81, 90
Numbers from 10 to 100 by 10	10, 20, 30, 40, 50, 60, 70, 80, 90, 100

Review 1 B

Addition the same numbers

$$48 + 48 = 40 + 8 + 40 + 8 = 40 + 40 + 8 + 8 = 80 + 16 = 96$$

	a	b	c	d
# 1	11+11	21+21	31+31	41+41
# 2	12+12	22+22	32+32	42+42
# 3	13+13	23+23	33+33	43+43
# 4	14+14	24+24	34+34	44+44
# 5	15+15	25+25	35+35	45+45
# 6	16+16	26+26	36+36	46+46
# 7	17+17	27+27	37+37	47+47
# 8	18+18	28+28	38+38	48+48
# 9	19+19	29+29	39+39	49+49
# 10	20+20	30+30	40+40	50+50

	a	b	c	d
# 11	21+21	31+31	41+41	51+51
# 12	22+22	32+32	42+42	52+52
# 13	23+23	33+33	43+43	53+53
# 14	24+24	34+34	44+44	54+54
# 15	25+25	35+35	45+45	55+55
# 16	26+26	36+36	46+46	56+56
# 17	27+27	37+37	47+47	57+57
# 18	28+28	38+38	48+48	58+58
# 19	29+29	39+39	49+49	59+59
# 20	30+30	40+40	50+50	60+60

Grade 4 Chapter 1 Lesson 3

Adding three numbers

Three numbers add in any order

11 + 14 + 19 = 11 + 19 + 14 = 30 + 14 = 44

11 + 14 + 19 = 25 + 19 = 44

11 + 14 + 19 = 11 + 33 = 44

Review 1 A

Add numbers

	a	b	c	d
# 1	20 + 0 + 9	30 + 0 + 9	40 + 0 + 9	50 + 0 + 9
# 2	21 + 1 + 8	31 + 1 + 8	41 + 1 + 8	51 + 1 + 8
# 3	22 + 2 + 7	32 + 2 + 7	42 + 2 + 7	52 + 2 + 7
# 4	23 + 3 + 6	33 + 3 + 6	43 + 3 + 6	53 + 3 + 6
# 5	24 + 4 + 5	34 + 4 + 5	44 + 4 + 5	54 + 4 + 5
# 6	25 + 5 + 4	35 + 5 + 4	45 + 5 + 4	55 + 5 + 4
# 7	26 + 6 + 3	36 + 6 + 3	46 + 6 + 3	56 + 6 + 3
# 8	27 + 7 + 2	37 + 7 + 2	47 + 7 + 2	57 + 7 + 2
# 9	28 + 8 + 1	38 + 8 + 1	48 + 8 + 1	58 + 8 + 1
# 10	29 + 9 + 0	39 + 9 + 0	49 + 9 + 0	59 + 9 + 0

Subtract three numbers

To subtract three numbers, work from left to right

19 – 1 – 9 = 18 – 9 = 9

18 – 2 – 8 = 16 – 8 = 8

18 – 8 – 2 = 10 – 2 = 8

Subtract numbers

	a	b	c	d
# 11	60 – 9 – 0	70 – 9 – 0	80 – 9 – 0	90 – 9 – 0
# 12	61 – 8 – 1	71 – 8 – 1	81 – 8 – 1	91 – 8 – 1
# 13	62 – 7 – 2	72 – 7 – 2	82 - 7 – 2	92 – 7 – 2
# 14	63 – 6 – 3	73 – 6 – 3	83 – 6 – 3	93 – 6 – 3
# 15	64 – 5 – 4	74 – 5 – 4	84 – 5 – 4	94 – 5 – 4
# 16	65 – 4 – 5	75 – 4 – 5	85 - 4 – 5	95 – 4 – 5
# 17	66 – 3 – 6	76 – 3 – 6	86 – 3 – 6	96 – 3 – 6
# 18	67 – 2 – 7	77 – 2 – 7	87 – 2 – 7	97 – 2 – 7
# 19	68 – 1 – 8	78 – 1 – 8	88 – 1 – 8	98 – 1 – 8
# 20	69 – 0 – 9	79 – 0 – 9	89 – 0 – 9	99 – 0 – 9

Grade 4 Chapter 1 Lesson 3
Two operations with numbers
Review 1 B

To do two operations with numbers work from left to right

19 + 7 - 1 = 26 - 1 = 25

18 + 8 - 2 = 26 - 2 = 24

Add and Subtract numbers

	a	b	c	d
# 1	20 + 9 – 0	30 + 9 – 0	40 + 9 – 0	50 + 9 - 0
# 2	21 + 8 – 1	31 + 8 – 1	41 + 8 – 1	51 + 8 - 1
# 3	22 + 7 – 2	32 + 7 – 2	42 + 7 – 2	52 + 7 - 2
# 4	23 + 6 – 3	33 + 6 – 3	43 + 6 – 3	53 + 6 - 3
# 5	24 + 5 – 4	34 + 5 – 4	44 + 5 – 4	54 + 5 - 4
# 6	25 + 4 – 5	35 + 4 – 5	45 + 4 – 5	55 + 4 - 5
# 7	26 + 3 – 6	36 + 3 – 6	46 + 3 – 6	56 + 3 - 6
# 8	27 + 2 – 7	37 + 2 – 7	47 + 2 – 7	57 + 2 - 7
# 9	28 + 1 – 8	38 + 1 – 8	48 + 1 – 8	58 + 1 - 8
# 10	29 + 0 – 9	39 + 0 – 9	49 + 0 – 9	59 + 0 - 9

Two operations with numbers

To do two operations with numbers work from left to right

19 - 8 + 9 = 11 + 9 = 20

18 - 9 + 8 = 9 + 8 = 17

Subtract and Add Numbers

	a	b	c	d
# 11	20 - 9 + 0	30 - 9 + 0	40 - 9 + 0	50 - 9 + 0
# 12	21 - 8 + 1	31 - 8 + 1	41 - 8 + 1	51 - 8 + 1
# 13	22 - 7 + 2	32 - 7 + 2	42 - 7 + 2	52 - 7 + 2
# 14	23 - 6 + 3	33 - 6 + 3	43 - 6 + 3	53 - 6 + 3
# 15	24 - 5 + 4	34 - 5 + 4	44 - 5 + 4	54 - 5 + 4
# 16	25 - 4 + 5	35 - 4 + 5	45 - 4 + 5	55 - 4 + 5
# 17	26 - 3 + 6	36 - 3 + 6	46 - 3 + 6	56 - 3 + 6
# 18	27 - 2 + 7	37 - 2 + 7	47 - 2 + 7	57 - 2 + 7
# 19	28 - 1 + 8	38 - 1 + 8	48 - 1 + 8	58 - 1 + 8
# 20	29 - 0 + 9	39 - 0 + 9	49 - 0 + 9	59 - 0 + 9

Grade 4 Chapter 1 Lesson 4
Addition and Subtraction
Three-digit numbers

Three-digit numbers content three digits
This is numbers 100, 101, 102, … 999

Least three-digit number: 100
Greatest three-digit number: 999

Digits for three-digit numbers from right to the left:
Ones, tens, and hundreds 238 = 200 + 30 + 8

One-digit numbers less than two or three-digit numbers
Three-digit numbers greater than one or two-digit numbers

To compare numbers, compare first hundreds, after that tens, then ones

$423 < 456$ $789 > 786$

Two numbers add in any order. Sum will be the same.
200 + 300 = 500   300 + 200 = 500

The sum of a number and zero is the same number
500 + 0 = 500   0 + 500 = 500

Difference any number and zero is the same number   500 – 0 = 500
Difference two the same number is zero   500 – 500 = 0

Adding three-digit numbers: from right to left
Add the ones. Add the tens. Add the hundreds.
Subtracting three-digit numbers: from right to left
Subtract the ones. Subtract the tens. Subtract the hundreds.

		1 1	7 14 14
246	678	154	8 5 4
+ 351	- 432	+789	- 2 8 6
597	246	943	5 6 8

Grade 4 Chapter 1 Lesson 4

Add three-digit numbers

	a	b	c	d	e
# 1	121 +594	132 +616	143 +715	154 +495	465 +484
# 2	252 +732	276 +660	312 +629	408 +588	561 +348
# 3	273 +598	286 +589	299 +592	377 +559	429 +546
# 4	294 +532	378 +476	392 +469	322 + 448	336 +439

Subtract three-digit numbers

	a	b	c	d	e
# 5	966 -357	945 -378	924 -399	903 -504	882 -525
# 6	989 -598	943 -575	827 -499	862 -483	815 -437
# 7	999 -108	955 -359	891 -297	837 -248	784 -189
# 8	957 -609	899 -493	841 -495	763 -377	718 -319

Grade 4 Chapter 1 Lesson 5

Sum from 100 to 199

Add numbers in column

	a	b	c	d	e
# 1	95 +22	96 +24	97 +26	98 +28	99 +29
# 2	38 +97	37 +96	36 +95	35 +99	34 +98
# 3	98 +43	97 +45	96 +47	95 + 49	97 +48
# 4	58 +97	57 +95	56 +98	55 +96	54 +99
# 5	97 +68	94 +66	99 +64	96 +65	95 +67
# 6	78 +97	77 +95	76 +98	75 +96	79 +94
# 7	96 +89	94 +86	99 +84	97 +85	95 +86
# 8	98 +97	97 +95	96 +94	95 +96	94 +99

Grade 4 Chapter 1 Lesson 5

Subtract numbers in column

	a	b	c	d	e
# 9	121 - 97	120 - 94	123 - 96	126 - 98	128 - 99
# 10	135 - 97	133 - 96	131 - 95	134 - 99	132 - 98
# 11	141 - 98	142 - 97	143 - 96	144 - 95	145 - 97
# 12	155 - 97	152 - 95	154 - 98	151 - 96	153 - 99
# 13	165 - 97	160 - 94	163 - 99	164 - 97	162 - 96
# 14	175 - 97	172 - 95	174 - 98	171 - 96	173 - 94
# 15	185 - 96	180 - 94	183 - 99	182 - 97	181 - 98
# 16	195 - 97	192 - 95	190 - 94	191 - 96	193 - 99

Grade 4 Chapter 1 Lesson 6

Sum from 200 to 299

Add numbers in column

	a	b	c	d	e
# 1	195 +22	196 +24	197 +26	198 +28	199 +29
# 2	138 +97	137 +96	136 +95	135 +99	134 +98
# 3	198 +43	197 +45	196 +47	195 +49	197 +48
# 4	158 +97	157 +95	156 +98	155 +96	154 +99
# 5	197 +68	194 +66	199 +64	196 +65	195 +67
# 6	178 +97	177 +95	176 +98	175 +96	179 +94
# 7	196 +89	194 +86	199 +84	197 +85	195 +86
# 8	198 +97	197 +95	196 +94	195 +96	194 +99

Grade 4 Chapter 1 Lesson 6

Subtract numbers in column

	a	b	c	d	e
# 9	221 - 97	220 - 94	223 - 96	226 - 98	228 - 99
# 10	235 - 97	233 - 96	231 - 95	234 - 99	232 - 98
# 11	241 - 98	242 - 97	243 - 96	244 - 95	245 - 97
# 12	255 - 97	252 - 95	254 - 98	251 - 96	253 - 99
# 13	265 - 97	260 - 94	263 - 99	264 - 97	262 - 96
# 14	275 - 97	272 - 95	274 - 98	271 - 96	273 - 94
# 15	285 - 96	280 - 94	283 - 99	282 - 97	281 - 98
# 16	295 - 97	292 - 95	290 - 94	291 - 96	293 - 99

Grade 4 Chapter 1 Lesson 7

Sum from 300 to 399

Add numbers in column

	a	b	c	d	e
# 1	195 +122	196 +124	197 +126	198 +128	199 +129
# 2	138 +197	137 +196	136 +195	135 +199	134 +198
# 3	198 +143	197 +145	196 +147	195 +149	197 +148
# 4	158 +197	157 +195	156 +198	155 +196	154 +199
# 5	197 +168	194 +166	199 +164	196 +165	195 +167
# 6	178 +197	177 +195	176 +198	175 +196	179 +194
# 7	196 +189	194 +186	199 +184	197 +185	195 +186
# 8	198 +197	197 +195	196 +194	195 +196	194 +199

Grade 4 Chapter 1 Lesson 7

Subtract numbers in column

	a	b	c	d	e
# 9	321 - 97	320 - 94	323 - 96	326 - 98	328 - 99
# 10	335 - 97	333 - 96	331 - 95	334 - 99	332 - 98
# 11	341 - 98	342 - 97	343 - 96	344 - 95	345 - 97
# 12	355 - 97	352 - 95	354 - 98	351 - 96	353 - 99
# 13	365 - 97	360 - 94	363 - 99	364 - 97	362 - 96
# 14	375 - 97	372 - 95	374 - 98	371 - 96	373 - 94
# 15	385 - 96	380 - 94	383 - 99	382 - 97	381 - 98
# 16	395 - 97	392 - 95	390 - 94	391 - 96	393 - 99

Grade 4 Chapter 1 Lesson 8
Addition numbers
To add numbers, work in any order

410 + 140 + 290 = 550 + 290 = 840

410 + 140 + 290 = 410 + 430 = 840

410 + 140 + 290=410 + 290 + 140=700 + 140=840

Evaluate

	a	b	c
# 1	10 + 90 + 10	20 + 80 + 20	30 + 70 + 30
# 2	40 + 60 + 40	50 + 50 + 50	60 + 40 + 60
# 3	70 + 30 + 70	80 + 20 + 80	90 + 10 + 90
# 4	100 + 0 +100	110 + 90 + 10	120 + 80 + 20
# 5	130 + 70 + 30	140 + 60 + 40	150 + 50 + 50
# 6	160 + 40 + 60	170 + 30 + 70	180 + 20 + 80
# 7	190 + 100 + 0	100+100+100	210 + 90 + 10
# 8	220 + 80 + 20	230 + 70 + 30	240 + 60 + 40
# 9	250 + 50 + 50	260 + 40 + 60	270 + 30 + 70
# 10	280 + 20 + 80	290 + 10 + 90	300 + 100 + 0

Add numbers in column

$$\begin{array}{r} 220 \\ +\ 150 \\ \underline{+\ 380} \\ 750 \end{array}$$

Evaluate

	a	b	c
# 11	310 + 90 + 290	320 + 80 + 280	330 + 70 + 270
# 12	340 + 60 + 260	350 + 50 + 250	360 + 40 + 240
# 13	370 + 30 + 230	380 + 20 + 220	390 + 10 + 210
# 14	400 + 20 + 300	410 + 30 + 290	420 + 40 + 280
# 15	430 + 50 + 270	440 + 60 + 260	450 + 70 + 250
# 16	460 + 80 + 240	470 + 90 + 230	480 + 100 + 220
# 17	490 + 110 + 210	500 + 120 + 200	510 + 130 + 190
# 18	520 + 140 + 180	530 + 150 + 170	540 + 160 + 160
# 19	550 + 170 + 150	560 + 180 + 140	570 + 190 + 130
# 20	580 + 200 + 120	590 + 210 + 110	600 + 220 + 100

Grade 4 Chapter 1 Lesson 8
Addition and Subtraction numbers
To add and subtract numbers work from left to right
410 + 140 - 110 = 550 - 110 = 440

Evaluate

	a	b	c
# 21	10 + 90 - 0	20 + 80 - 10	30 + 70 - 20
# 22	40 + 60 - 30	50 + 50 - 40	60 + 40 - 50
# 23	70 + 30 - 60	80 + 20 - 70	90 + 10 - 80
# 24	100 + 100 - 0	110 + 90 - 10	120 + 80 - 20
# 25	130 + 70 - 30	140 + 60 - 40	150 + 50 - 50
# 26	160 + 40 - 60	170 + 30 - 70	180 + 20 - 80
# 27	290 + 110 - 90	200 + 100 - 0	210 + 90 - 10
# 28	220 + 80 - 20	230 + 70 - 30	240 + 60 - 40
# 29	250 + 50 - 50	260 + 40 - 60	270 + 30 - 70
# 30	280 + 20 - 80	290 + 10 - 90	300 + 0 - 100

Add and subtract numbers in column
410 + 140 - 110 = 440

$$1)\quad \begin{array}{r} 410 \\ +\,140 \\ \hline 550 \end{array} \qquad 2)\quad \begin{array}{r} 550 \\ -\,110 \\ \hline 440 \end{array}$$

Evaluate

	a	b	c
# 31	310 + 90 - 10	320 + 80 - 20	330 + 70 - 30
# 32	340 + 60 - 40	350 + 50 - 50	360 + 40 - 60
# 33	370 + 30 - 70	380 + 20 - 80	390 + 10 - 90
# 34	400 + 300 - 0	410 + 290 - 10	420 + 280 - 20
# 35	430 + 270 - 30	440 + 260 - 40	450 + 250 - 50
# 36	460 + 240 - 60	470 + 230 - 70	480 + 220 - 80
# 37	490 + 310 - 190	500 + 200 - 100	510 + 190 - 110
# 38	520 + 180 - 120	530 + 170 - 130	540 + 160 - 140
# 39	550 + 150 - 150	560 + 140 - 160	570 + 130 - 170
# 40	580 + 120 - 180	590 + 110 - 190	600 + 100 - 200

Grade 4 Chapter 1 Lesson 9
Subtraction and Addition numbers
To subtract and add numbers work from left to right
410 - 110 + 140 = 300 + 140 = 440
Evaluate

	a	b	c
# 1	10 – 0 + 90	20 – 10 + 80	30 – 20 + 70
# 2	40 – 30 + 60	50 – 40 + 50	60 – 50 + 40
# 3	70 – 60 + 30	80 – 70 + 20	90 – 80 + 10
# 4	100 – 0 + 100	110 – 10 + 90	120 – 20 + 80
# 5	130 – 30 + 70	140 – 40 + 60	150 - 50 + 50
# 6	160 – 60 + 40	170 – 70 + 30	180 – 80 + 20
# 7	290 – 90 + 110	200 – 0 + 100	210 – 10 + 90
# 8	220 – 20 + 80	230 – 30 + 70	240 – 40 + 60
# 9	250 - 50 + 50	260 – 60 + 40	270 – 70 + 30
# 10	280 – 80 + 20	290 – 90 + 10	300 – 100 + 0

Subtract and add numbers in column
410 - 110 + 140 = 440

1) $\begin{array}{r} 410 \\ \underline{-\ 110} \\ 300 \end{array}$ 2) $\begin{array}{r} 300 \\ \underline{+\ 140} \\ 440 \end{array}$

Evaluate

	a	b	c
# 11	310 – 10 + 90	320 – 20 + 80	330 – 30 + 70
# 12	340 – 40 + 60	350 - 50 + 50	360 – 60 + 40
# 13	370 – 70 + 30	380 – 80 + 20	390 – 90 + 10
# 14	400 – 0 + 300	410 – 10 + 290	420 – 20 + 280
# 15	430 – 30 + 270	440 – 40 + 260	450 – 50 + 250
# 16	460 – 60 + 240	470 – 70 + 230	480 – 80 + 220
# 17	490 – 190 + 310	500 – 100 + 200	510 – 110 + 190
# 18	520 – 120 + 180	530 – 130 + 170	540 – 140 + 160
# 19	550 - 150 + 150	560 – 160 + 140	570 – 170 + 130
# 20	580 – 180 + 120	590 – 190 + 110	600 – 200 + 100

Grade 4 Chapter 1 Lesson 9
Subtraction numbers
To subtract numbers, work from left to right
900 – 200 – 100 = 700 – 100 = 600

Evaluate

	a	b	c
# 21	400 – 10 - 100	410 – 20 - 110	420 – 30 - 120
# 22	430 – 40 - 130	440 – 50 – 140	450 – 60 - 150
# 23	460 – 70 – 160	470 - 80 – 170	480 – 90 - 180
# 24	490 – 100 – 190	500 – 110 – 200	510 – 120 - 210
# 25	520 – 130 – 220	530 – 140 – 230	540 – 150 - 240
# 26	550 – 160 - 250	560 – 170 – 260	570 - 180 - 270
# 27	580 – 190 – 280	590 – 200 – 290	600 – 210 - 300
# 28	610 – 220 – 310	620 – 230 – 320	630 – 240 - 330
# 29	640 – 250 – 340	650 – 260 – 350	660 – 270 - 360
# 30	670 – 280 – 370	680 – 290 – 380	690 – 300 - 390

Subtract numbers in column
900 – 200 – 100 = 600

$$1)\quad \begin{array}{r} 900 \\ \underline{-\,200} \\ 700 \end{array} \qquad\qquad 2)\quad \begin{array}{r} 700 \\ \underline{-\,100} \\ 600 \end{array}$$

Evaluate

	a	b	c
# 31	700 – 10 – 400	710 – 20 – 410	720 – 30 - 420
# 32	730 – 40 – 430	740 – 50 – 440	750 – 60 - 450
# 33	760 – 70 – 460	770 – 80 – 470	780 – 90 - 480
# 34	790 – 100 – 490	800 – 110 – 500	810 – 120 - 510
# 35	820 – 130 – 520	830 – 140 – 530	840 – 150 - 540
# 36	850 – 160 – 550	860 - 170 – 560	870 – 180 - 570
# 37	880 – 190 – 580	890 – 200 – 590	900 – 210 - 600
# 38	910 – 220 – 610	920 – 230 – 620	930 – 240 - 630
# 39	940 – 250 – 640	950 – 260 – 650	960 – 270 - 660
# 40	970 – 280 – 670	980 – 290 – 680	990 – 300 - 690

Grade 4 Chapter 1 Lesson 10
Test 1 A
Add numbers in column

	a	b	c
# 1	196 +189	194 +186	199 +184

Subtract numbers in column

	a	b	c
# 2	385 - 96	380 - 94	383 - 99

Evaluate in row or in column

	a	b	c
# 3	250 + 50 + 50	260 + 40 + 60	270 + 30 + 70
# 4	550 + 170 + 150	560 + 180 + 140	570 + 190 + 130
# 5	250 + 50 - 50	260 + 40 - 60	270 + 30 - 70
# 6	550 + 150 - 150	560 + 140 - 160	570 + 130 – 170

	a	b	c
# 7	250 - 50 + 50	260 – 60 + 40	270 – 70 + 30
# 8	550 - 150 + 150	560 – 160 + 140	570 – 170 + 130
# 9	640 – 250 – 340	650 – 260 – 350	660 – 270 - 360
# 10	940 – 250 – 640	950 – 260 – 650	960 – 270 - 660

Grade 4 Chapter 1 Lesson 10
Test 1 B

Add numbers in column

	a	b	c
# 1	198 +197	197 +195	196 +194

Subtract numbers in column

	a	b	c
# 2	395 - 97	392 - 95	390 - 94

Evaluate in row or in column

	a	b	c
# 3	280 + 20 + 80	290 + 10 + 90	300 + 100 + 0
# 4	580 + 200 + 120	590 + 210 + 110	600 + 220 + 100
# 5	280 + 20 - 80	290 + 10 - 90	300 + 0 - 100
# 6	580 + 120 - 180	590 + 110 - 190	600 + 100 - 200

	a	b	c
# 7	280 – 80 + 20	290 – 90 + 10	300 – 100 + 0
# 8	580 – 180 + 120	590 – 190 + 110	600 – 200 + 100
# 9	670 – 280 – 370	680 – 290 – 380	690 – 300 - 390
# 10	970 – 280 – 670	980 – 290 – 680	990 – 300 - 690

Grade 4 Answers Chapter 1

Lesson 1

# 1	30	# 11	8	# 21	40	# 31	13
# 2	31	# 12	19	# 22	42	# 32	14
# 3	34	# 13	7	# 23	44	# 33	29
# 4	35	# 14	16	# 24	46	# 34	28
# 5	36	# 15	27	# 25	48	# 35	34
# 6	33	# 16	19	# 26	31	# 36	25
# 7	32	# 17	8	# 27	43	# 37	16
# 8	39	# 18	11	# 28	45	# 38	27
# 9	38	# 19	19	# 29	47	# 39	28
# 10	37	# 20	19	# 30	49	# 40	19

Lesson 2 Review 1 A

	a	b	c	d		a	b	c	d
# 1	+	+	+	+	# 11	+	+	+	+
# 2	-	-	-	-	# 12	-	-	-	-
# 3	+	+	+	+	# 13	+	+	+	+
# 4	-	-	-	-	# 14	-	-	-	-
# 5	+	+	+	+	# 15	+	+	+	+
# 6	-	-	-	-	# 16	-	-	-	-
# 7	+	+	+	+	# 17	+	+	+	+
# 8	-	-	-	-	# 18	-	-	-	-
# 9	+	+	+	+	# 19	+	+	+	+
# 10	-	-	-	-	# 20	-	-	-	-

Lesson 2 Review 1 B

	a b c d		a	b	c	d
# 1	22, 42, 62, 82	# 11	42,	62,	82,	102
# 2	24, 44, 64, 84	# 12	44,	64,	84,	104
# 3	26, 46, 66, 86	# 13	46,	66,	86,	106
# 4	28, 48, 68, 88	# 14	48,	68,	88,	108
# 5	30, 50, 70, 90	# 15	50,	70,	90,	110
# 6	32, 52, 72, 92	# 16	52,	72,	92,	112
# 7	34, 54, 74, 94	# 17	54,	74,	94,	114
# 8	36, 56, 76, 96	# 18	56,	76,	96,	116
# 9	38, 58, 78, 98	# 19	58,	78,	98,	118
# 10	40, 60, 80, 100	# 20	60,	80,	100,	120

Grade 4 Answers Chapter 1

Lesson 3 Review 1 A

	a b c d		a b c d
# 1	29, 39, 49, 59	# 11	51, 61, 71, 81
# 2	30, 40, 50, 60	# 12	52, 62, 72, 82
# 3	31, 41, 51, 61	# 13	53, 63, 73, 83
# 4	32, 42, 52, 62	# 14	54, 64, 74, 84
# 5	33, 43, 53, 63	# 15	55, 65, 75, 85
# 6	34, 44, 54, 64	# 16	56, 66, 76, 86
# 7	35, 45, 55, 65	# 17	57, 67, 77, 87
# 8	36, 46, 56, 66	# 18	58, 68, 78, 88
# 9	37, 47, 57, 67	# 19	59, 69, 79, 89
# 10	38, 48, 58, 68	# 20	60, 70, 80, 90

Lesson 3 Review 1 B

	a b c d		a b c d
# 1	29, 39, 49, 59	# 11	11, 21, 31, 41
# 2	28, 38, 48, 58	# 12	14, 24, 34, 44
# 3	27, 37, 47, 57	# 13	17, 27, 37, 47
# 4	26, 36, 46, 56	# 14	20, 30, 40, 50
# 5	25, 35, 45, 55	# 15	23, 33, 43, 53
# 6	24, 34, 44, 54	# 16	26, 36, 46, 56
# 7	23, 33, 43, 53	# 17	29, 39, 49, 59
# 8	22, 32, 42, 52	# 18	32, 42, 52, 62
# 9	21, 31, 41, 51	# 19	35, 45, 55, 65
# 10	20, 30, 40, 50	# 20	38, 48, 58, 68

Lesson 4	a b c d e		a b c d e
# 1	715, 748, 858, 649, 949	# 5	609, 567, 525, 399, 357
# 2	984, 936, 941, 996, 909	# 6	391, 368, 328, 379, 378
# 3	871, 875, 891, 936, 975	# 7	891, 596, 594, 589, 595
# 4	826, 854, 861, 770, 775	# 8	348, 406, 346, 386, 399

Lesson 5	a b c d e		a b c d e
# 1	117, 120, 123, 126, 128	# 9	24, 26, 27, 28, 29
# 2	135, 133, 131, 134, 132	# 10	38, 37, 36, 35, 34
# 3	141, 142, 143, 144, 145	# 11	43, 45, 47, 49, 48
# 4	155, 152, 154, 151, 153	# 12	58, 57, 56, 55, 54
# 5	165, 160, 163, 161, 162	# 13	68, 66, 64, 67, 66
# 6	175, 172, 174, 171, 173	# 14	78, 77, 76, 75, 79
# 7	185, 180, 183, 182, 181	# 15	89, 86, 84, 85, 83
# 8	195, 192, 190, 191, 193	# 16	98, 97, 96, 95, 94

Grade 4 Answers Chapter 1

Lesson 6	a	b	c	d	e
# 1	217	220	223	226	228
# 2	235	233	231	234	232
# 3	241	242	243	244	245
# 4	255	252	254	251	253
# 5	265	260	263	261	262
# 6	275	272	274	271	273
# 7	285	280	283	282	281
# 8	295	292	290	291	293
# 9	124	126	127	128	129
# 10	138	137	136	135	134
# 11	143	145	147	149	148
# 12	158	157	156	155	154
# 13	168	166	164	167	166
# 14	178	177	176	175	179
# 15	189	186	184	185	183
# 16	198	197	196	195	194

Lesson 7	a	b	c	d	e
# 1	317	320	323	326	328
# 2	335	333	331	334	332
# 3	341	342	343	344	345
# 4	355	352	354	351	353
# 5	365	360	363	361	362
# 6	375	372	374	371	373
# 7	385	380	383	382	381
# 8	395	392	390	391	393
# 9	224	226	227	228	229
# 10	238	237	236	235	234
# 11	243	245	247	249	248
# 12	258	257	256	255	254
# 13	268	266	264	267	266
# 14	278	277	276	275	279
# 15	289	286	284	285	283
# 16	298	297	296	295	294

Lesson 8	a	b	c
# 1	110	120	130
# 2	140	150	160
# 3	170	180	190
# 4	200	210	220
# 5	230	240	250
# 6	260	270	280
# 7	290	300	310
# 8	320	330	340
# 9	350	360	370
# 10	380	390	400
# 11	690	680	670
# 12	660	650	640
# 13	630	620	610
# 14	720	730	740
# 15	750	760	770
# 16	780	790	800
# 17	810	820	830
# 18	840	850	860
# 19	870	880	890
# 20	900	910	920
# 21	100	90	80
# 22	70	60	50
# 23	40	30	20
# 24	200	190	180
# 25	170	160	150
# 26	140	130	120
# 27	310	300	290
# 28	280	270	260
# 29	250	240	230
# 30	220	210	200
# 31	390	380	370
# 32	360	350	340
# 33	330	320	310
# 34	700	690	680
# 35	670	660	650
# 36	640	630	620
# 37	610	600	590
# 38	580	570	560
# 39	550	540	530
# 40	520	510	500

Grade 4 Answers Chapter 1

Lesson 9

	a	b	c
# 1	100	90	80
# 2	70	60	50
# 3	40	30	20
# 4	200	190	180
# 5	170	160	150
# 6	140	130	120
# 7	310	300	290
# 8	280	270	260
# 9	250	240	230
# 10	220	210	200

	a	b	c
# 11	390	380	370
# 12	360	350	340
# 13	330	320	310
# 14	700	690	680
# 15	670	660	650
# 16	640	630	620
# 17	610	600	590
# 18	580	570	560
# 19	550	540	530
# 20	520	510	500

Lesson 9

	a	b	c
# 21	290	280	270
# 22	260	250	240
# 23	230	220	210
# 24	200	190	180
# 25	170	160	150
# 26	140	130	120
# 27	110	100	90
# 28	80	70	60
# 29	50	40	30
# 30	20	10	0

	a	b	c
# 31	290	280	270
# 32	260	250	240
# 33	230	220	210
# 34	200	190	180
# 35	170	160	150
# 36	140	130	120
# 37	110	100	90
# 38	80	70	60
# 39	50	40	30
# 40	20	10	0

Lesson 10 Test 1 A

	a	b	c
# 1	385	380	383
# 2	289	286	284
# 3	350	360	370
# 4	870	880	890
# 5	250	240	230
# 6	550	540	530
# 7	250	240	230
# 8	550	540	530
# 9	50	40	30
# 10	50	40	30

Test 1 B

	a	b	c
# 1	395	392	390
# 2	298	297	296
# 3	380	390	400
# 4	900	910	920
# 5	220	210	200
# 6	520	510	500
# 7	220	210	200
# 8	520	510	500
# 9	20	10	0
# 10	20	10	0

Grade 4

Chapter 2
Multiplication
Numbers

Grade 4 Chapter 2 Contents
Multiplication Numbers
Lesson 1 – 14
Basic Problems

Work with lesson 1, read rules and examples, 2) Solve the problems from contents, 3) Check answers in the end of chapter 1, 4) Work with the next lessons 2, 3, 4, 5, …, 5) Solve Test 1 A, 6) Grading Scale: A =10 points, B=9points, C=8 points, D=7 points, F=0-6 points, 7) If you solve 10 problems correctly, you get an A, 8) If have an A, you can move on to next chapter, 9) If your grade is lower than an A, work with the mistakes, then solve Test 1B, 10) When you finish all of the chapters, solve Final Exam in last chapter.

Grade 4 Chapter 2 Lesson 1
Multiplication one-digit numbers

Third operation with numbers is multiplication × or ·
Multiplication is the operation of repeated addition
$2\times 3=2+2+2=6$ or $2\cdot 3=2+2+2=6$

First number in multiplication called the factor	**2**
Second number in multiplication called the factor	**3**
Result of multiplication called the product	**6**

Properties of multiplication

Two numbers can multiply in any order. The product will be the same
$2\times 3=2+2+2=6$ $3\times 2=3+3=6$
The product of a number and zero is zero: $5\times 0=0$ $0\times 5=0$
The product of a number and one is the same number: $5\times 1=5$ $1\times 5=5$
Find the product: $4\times 8=8\times 4=8+8+8+8=16+16=32$

	a	b
# 1	$0\times 2=0+0=0$	$0\times 3=0+0+0=0$
# 2	$1\times 2=1+1=2$	$1\times 3=1+1+1=3$
# 3	$2\times 2=2+2=4$	$2\times 3=2+2+2=6$
# 4	$3\times 2=3+3=6$	$3\times 3=3+3+3=9$
# 5	$4\times 2=4+4=8$	$4\times 3=4+4+4=12$
# 6	$5\times 2=5+5=10$	$5\times 3=5+5+5=15$
# 7	$6\times 2=6+6=12$	$6\times 3=6+6+6=18$
# 8	$7\times 2=7+7=14$	$7\times 3=7+7+7=21$
# 9	$8\times 2=8+8=16$	$8\times 3=8+8+8=24$
# 10	$9\times 2=9+9=18$	$9\times 3=9+9+9=27$

	a	b		a	b
# 11	$0\times 4=0$	$0\times 5=0$	# 21	$6\times 0=0$	$7\times 0=0$
# 12	$1\times 4=4$	$1\times 5=5$	# 22	$6\times 1=6$	$7\times 1=7$
# 13	$2\times 4=8$	$2\times 5=10$	# 23	$6\times 2=12$	$7\times 2=14$
# 14	$3\times 4=12$	$3\times 5=15$	# 24	$6\times 3=18$	$7\times 3=21$
# 15	$4\times 4=16$	$4\times 5=20$	# 25	$6\times 4=24$	$7\times 4=28$
# 16	$5\times 4=20$	$5\times 5=25$	# 26	$6\times 5=30$	$7\times 5=35$
# 17	$6\times 4=24$	$6\times 5=30$	# 27	$6\times 6=36$	$7\times 6=42$
# 18	$7\times 4=28$	$7\times 5=35$	# 28	$6\times 7=42$	$7\times 7=49$
# 19	$8\times 4=32$	$8\times 5=40$	# 29	$6\times 8=48$	$7\times 8=56$
# 20	$9\times 4=36$	$9\times 5=45$	# 30	$6\times 9=54$	$7\times 9=63$

Grade 4 Chapter 2 Lesson 2

Multiply the numbers and memorize the result:

$2\times3=2+2+2=6$ $3\times2=3+3=6$ $3\times9=9\times3=9+9+9=18+9=27$

	a	b	c	d
# 1	2×1	3×1	4×1	5×1
# 2	2×2	3×2	4×2	5×2
# 3	2×3	3×3	4×3	5×3
# 4	2×4	3×4	4×4	5×4
# 5	2×5	3×5	4×5	5×5
# 6	2×6	3×6	4×6	5×6
# 7	2×7	3×7	4×7	5×7
# 8	2×8	3×8	4×8	5×8
# 9	2×9	3×9	4×9	5×9
# 10	2×10	3×10	4×10	5×10
#11	6×1	7×1	8×1	9×1
# 12	6×2	7×2	8×2	9×2
# 13	6×3	7×3	8×3	9×3
# 14	6×4	7×4	8×4	9×4
# 15	6×5	7×5	8×5	9×5
# 16	6×6	7×6	8×6	9×6
# 17	6×7	7×7	8×7	9×7
# 18	6×8	7×8	8×8	9× 8
# 19	6×9	7×9	8×9	9×9
# 20	6×10	7×10	8×10	9×10

Table of multiplication

Memorize the Table of Multiplication. Number 5 from the left column multiply by number 4 from first row to get, answer 20

	×	2	3	4	5	6	7	8	9
# 1	1	2	3	4	5	6	7	8	9
# 2	2	4	6	8	10	12	14	16	18
# 3	3	6	9	12	15	18	21	24	27
# 4	4	8	12	16	20	24	28	32	36
# 5	5	10	15	20	25	30	35	40	45
# 6	6	12	18	24	30	36	42	48	54
# 7	7	14	21	28	35	42	49	56	63
# 8	8	16	24	32	40	48	56	64	72
# 9	9	18	27	36	45	54	63	72	81
# 10	10	20	30	40	50	60	70	80	90

Grade 4 Chapter 2 Lesson 3
Multiplication numbers from 10 to 19
Multiply numbers in row

$5 \times 10 = 50$ $10 \times 5 = 50$ $5 \times 11 = 55$ $11 \times 5 = 55$

1) 15×9
2) $15 = 10 + 5$
3) $10 \times 9 = 90$
4) $5 \times 9 = 45$
5) $90 + 45 = 135$
6) $15 \times 9 = 135$

Multiply numbers

	a	b	c	d	e
# 1	10×2	11×2	12×2	13×2	14×2
# 2	10×3	11×3	12×3	13×3	14×3
# 3	10×4	11×4	12×4	13×4	14×4
# 4	10×5	11×5	12×5	13×5	14×5
# 5	10×6	11×6	12×6	13×6	14×6
# 6	10×7	11×7	12×7	13×7	14×7
# 7	10×8	11×8	12×8	13×8	14×8
# 8	10×9	11×9	12×9	13×9	14×9
# 9	15×2	16×2	17×2	18×2	19×2
# 10	15×3	16×3	17×3	18×3	19×3

Grade 4 Chapter 2 Lesson 3
Multiply numbers in column

$$\begin{array}{r} 15 \\ \underline{\times 9} \\ 45 \\ \underline{+\,90} \\ 135 \end{array}$$

1) $9 \times 5 = 45$
2) $9 \times 10 = 90$

Multiply numbers

	a	b	c	d	e
# 11	15×4	16×4	17×4	18×4	19×4
# 12	15×5	16×5	17×5	18×5	19×5
# 13	15×6	16×6	17×6	18×6	19×6
# 14	15×7	16×7	17×7	18×7	19×7
# 15	15×8	16×8	17×8	18×8	19×8
# 16	15×9	16×9	17×9	18×9	19×9

Grade 4 Chapter 2 Lesson 4
Multiplication numbers from 20 to 29

Multiply numbers in row

1) 24×9
2) $24 = 20 + 4$
3) $20 \times 9 = 180$
4) $4 \times 9 = 36$
5) $180 + 36 = 216$
6) $24\times 9 = 216$

Multiply numbers

	a	b	c	d	e
# 1	20×2	21×2	22×2	23×2	24×2
# 2	20×3	21×3	22×3	23×3	24×3
# 3	20×4	21×4	22×4	23×4	24×4
# 4	20×5	21×5	22×5	23×5	24×5
# 5	20×6	21×6	22×6	23×6	24×6
# 6	20×7	21×7	22×7	23×7	24×7
# 7	20×8	21×8	22×8	23×8	24×8
# 8	20×9	21×9	22×9	23×9	24×9
# 9	25×2	26×2	27×2	28×2	29×2
#10	25×3	26×3	27×3	28×3	29×3

Grade 4 Chapter 2 Lesson 4
Multiply numbers in column

6	1) $9 \times 7 = 63$
27	2) Put 3 below 9
$\underline{\times 9}$	3) Put 6 above 2
243	4) $9 \times 2 = 18$
	5) $18 + 6 = 24$
	6) Put 24 ahead 3

Multiply numbers

	a	b	c	d	e
# 11	25 $\times \underline{4}$	26 $\times \underline{4}$	27 $\times \underline{4}$	28 $\times \underline{4}$	29 $\times \underline{4}$
# 12	25 $\times \underline{5}$	26 $\times \underline{5}$	27 $\times \underline{5}$	28 $\times \underline{5}$	29 $\times \underline{5}$
# 13	25 $\times \underline{6}$	26 $\times \underline{6}$	27 $\times \underline{6}$	28 $\times \underline{6}$	29 $\times \underline{6}$
# 14	25 $\times \underline{7}$	26 $\times \underline{7}$	27 $\times \underline{7}$	28 $\times \underline{7}$	29 $\times \underline{7}$
# 15	25 $\times \underline{8}$	26 $\times \underline{8}$	27 $\times \underline{8}$	28 $\times \underline{8}$	29 $\times \underline{8}$
# 16	25 $\times \underline{9}$	26 $\times \underline{9}$	27 $\times \underline{9}$	28 $\times \underline{9}$	29 $\times \underline{9}$

Grade 4 Chapter 2 Lesson 5
Multiplication numbers from 30 to 39
Multiply numbers in row

1) 35×9
2) $35 = 30 + 5$
3) $30 \times 9 = 270$
4) $5 \times 9 = 45$
5) $270 + 45 = 315$
6) $35\times 9 = 315$

Multiply numbers

	a	b	c	d	e
# 1	30×2	31×2	32×2	33×2	34×2
# 2	30×3	31×3	32×3	33×3	34×3
# 3	30×4	31×4	32×4	33×4	34×4
# 4	30×5	31×5	32×5	33×5	34×5
# 5	30×6	31×6	32×6	33×6	34×6
# 6	30×7	31×7	32×7	33×7	34×7
# 7	30×8	31×8	32×8	33×8	34×8
# 8	30×9	31×9	32×9	33×9	34×9
# 9	35×2	36×2	37×2	38×2	39×2
# 10	35×3	36×3	37×3	38×3	39×3

Grade 4 Chapter 2 Lesson 5
Multiply numbers in column

$$\begin{array}{r} {}^{7}\\ 38 \\ \times\underline{9} \\ 342 \end{array}$$

1) $9 \times 8 = 72$
2) Put 2 below 9
3) Put 7 above 3
4) $9 \times 3 = 27$
5) $27 + 7 = 34$
6) Put 34 ahead 2

Multiply numbers

	a	b	c	d	e
# 11	35×4	36×4	37×4	38×4	39×4
# 12	35×5	36×5	37×5	38×5	39×5
# 13	35×6	36×6	37×6	38×6	39×6
# 14	35×7	36×7	37×7	38×7	39×7
# 15	35×8	36×8	37×8	38×8	39×8
# 16	35×9	36×9	37×9	38×9	39×9

Grade 4 Chapter 2 Lesson 6
Multiplication numbers from 40 to 49
Multiply numbers in row

1) 43×9
2) $43 = 40 + 3$
3) $40 \times 9 = 360$
4) $3 \times 9 = 27$
5) $360 + 27 = 387$
6) $43\times 9 = 387$

Multiply numbers

	a	b	c	d	e
# 1	40×2	41×2	42×2	43×2	44×2
# 2	40×3	41×3	42×3	43×3	44×3
# 3	40×4	41×4	42×4	43×4	44×4
# 4	40×5	41×5	42×5	43×5	44×5
# 5	40×6	41×6	42×6	43×6	44×6
# 6	40×7	41×7	42×7	43×7	44×7
# 7	40×8	41×8	42×8	43×8	44×8
# 8	40×9	41×9	42×9	43×9	44×9
# 9	45×2	46×2	47×2	48×2	49×2
# 10	45×3	46×3	47×3	48×3	49×3

Grade 4 Chapter 2 Lesson 6
Multiply numbers in column

$$\begin{array}{r} {}^{2} \\ 43 \\ \times \underline{9} \\ 387 \end{array}$$

1) $9 \times 3 = 27$
2) Put 7 below 9
3) Put 2 above 4
4) $9 \times 4 = 36$
5) $36 + 2 = 38$
6) Put 38 ahead 7

Multiply numbers

	a	b	c	d	e
# 11	45×4	46×4	47×4	48×4	49×4
# 12	45×5	46×5	47×5	48×5	49×5
# 13	45×6	46×6	47×6	48×6	49×6
# 14	45×7	46×7	47×7	48×7	49×7
# 15	45×8	46×8	47×8	48×8	49×8
# 16	45×9	46×9	47×9	48×9	49×9

Grade 4 Chapter 2 Lesson 7
Multiplication from 50 to 59
Multiply numbers in row

1) 57×9
2) $57 = 50 + 7$
3) $50 \times 9 = 450$
4) $7 \times 9 = 63$
5) $450 + 63 = 513$
6) $57\times 9 = 513$

Multiply numbers

	a	b	c	d	e
# 1	50×2	51×2	52×2	53×2	54×2
# 2	50×3	51×3	52×3	53×3	54×3
# 3	50×4	51×4	52×4	53×4	54×4
# 4	50×5	51×5	52×5	53×5	54×5
# 5	50×6	51×6	52×6	53×6	54×6
# 6	50×7	51×7	52×7	53×7	54×7
# 7	50×8	51×8	52×8	53×8	54×8
# 8	50×9	51×9	52×9	53×9	54×9
# 9	55×2	56×2	57×2	58×2	59×2
# 10	55×3	56×3	57×3	58×3	59×3

Grade 4 Chapter 2 Lesson 7
Multiply numbers in column

$$\begin{array}{r} {}^{6} \\ 57 \\ \times \underline{9} \\ 513 \end{array}$$

1) 9 × 7 = 63
2) Put 3 below 9
3) Put 6 above 5
4) 9 × 5 = 45
5) 45 + 6 = 51
6) Put 51 ahead 3

Multiply numbers

	a	b	c	d	e
# 11	$55 \times \underline{4}$	$56 \times \underline{4}$	$57 \times \underline{4}$	$58 \times \underline{4}$	$59 \times \underline{4}$
# 12	$55 \times \underline{5}$	$56 \times \underline{5}$	$57 \times \underline{5}$	$58 \times \underline{5}$	$59 \times \underline{5}$
# 13	$55 \times \underline{6}$	$56 \times \underline{6}$	$57 \times \underline{6}$	$58 \times \underline{6}$	$59 \times \underline{6}$
# 14	$55 \times \underline{7}$	$56 \times \underline{7}$	$57 \times \underline{7}$	$58 \times \underline{7}$	$59 \times \underline{7}$
# 15	$55 \times \underline{8}$	$56 \times \underline{8}$	$57 \times \underline{8}$	$58 \times \underline{8}$	$59 \times \underline{8}$
# 16	$55 \times \underline{9}$	$56 \times \underline{9}$	$57 \times \underline{9}$	$58 \times \underline{9}$	$59 \times \underline{9}$

Grade 4 Chapter 2 Lesson 8
Multiplication from 60 to 69
Multiply numbers in row

1) 63×9
2) $63 = 60 + 3$
3) $60 \times 9 = 540$
4) $3 \times 9 = 27$
5) $540 + 27 = 567$
6) $63\times 9 = 567$

Multiply numbers

	a	b	c	d	e
# 1	60×2	61×2	62×2	63×2	64×2
# 2	60×3	61×3	62×3	63×3	64×3
# 3	60×4	61×4	62×4	63×4	64×4
# 4	60×5	61×5	62×5	63×5	64×5
# 5	60×6	61×6	62×6	63×6	64×6
# 6	60×7	61×7	62×7	63×7	64×7
# 7	60×8	61×8	62×8	63×8	64×8
# 8	60×9	61×9	62×9	63×9	64×9
# 9	65×2	66×2	67×2	68×2	69×2
# 10	65×3	66×3	67×3	68×3	69×3

Grade 4 Chapter 2 Lesson 8
Multiply numbers in column

2	1) $9 \times 3 = 27$
63	2) Put 7 below 9
$\times \underline{9}$	3) Put 2 above 6
567	4) $9 \times 6 = 54$
	5) $54 + 2 = 56$
	6) Put 56 ahead 7

Multiply numbers

	a	b	c	d	e
# 11	65 ×4	66 ×4	67 ×4	68 ×4	69 ×4
# 12	65 ×5	66 ×5	67 ×5	68 ×5	69 ×5
# 13	65 ×6	66 ×6	67 ×6	68 ×6	69 ×6
# 14	65 ×7	66 ×7	67 ×7	68 ×7	69 ×7
# 15	65 ×8	66 ×8	67 ×8	68 ×8	69 ×8
# 16	65 ×9	66 ×9	67 ×9	68 ×9	69 ×9

Grade 4 Chapter 2 Lesson 9
Multiplication numbers from 70 – 79
Multiply numbers in row

1) 73×9
2) $73 = 70 + 3$
3) $70 \times 9 = 630$
4) $3 \times 9 = 27$
5) $630 + 27 = 657$
6) $73 \times 9 = 657$

Multiply numbers

	a	b	c	d	e
# 1	70×2	71×2	72×2	73×2	74×2
# 2	70×3	71×3	72×3	73×3	74×3
# 3	70×4	71×4	72×4	73×4	74×4
# 4	70×5	71×5	72×5	73×5	74×5
# 5	70×6	71×6	72×6	73×6	74×6
# 6	70×7	71×7	72×7	73×7	74×7
# 7	70×8	71×8	72×8	73×8	74×8
# 8	70×9	71×9	72×9	73×9	74×9
# 9	75×2	76×2	77×2	78×2	79×2
# 10	75×3	76×3	77×3	78×3	79×3

Grade 4 Chapter 2 Lesson 9
Multiply numbers in column

$$\begin{array}{r} {\scriptstyle 2}\\ 73 \\ \times\, \underline{9} \\ 657 \end{array}$$

1) $9 \times 3 = 27$
2) Put 7 below 9
3) Put 2 above 7
4) $9 \times 7 = 63$
5) $63 + 2 = 65$
6) Put 65 ahead 7

Multiply numbers

	a	b	c	d	e
# 11	75×4	76×4	77×4	78×4	79×4
# 12	75×5	76×5	77×5	78×5	79×5
# 13	75×6	76×6	77×6	78×6	79×6
# 14	75×7	76×7	77×7	78×7	79×7
# 15	75×8	76×8	77×8	78×8	79×8
# 16	75×9	76×9	77×9	78×9	79×9

Grade 4 Chapter 2 Lesson 10
Multiplication numbers from 80 to 89
Multiply numbers in row

1) 87×9
2) $87 = 80 + 7$
3) $80 \times 9 = 720$
4) $7 \times 9 = 63$
5) $720 + 63 = 783$
6) $87 \times 9 = 783$

Multiply numbers

	a	b	c	d	e
# 1	80×2	81×2	82×2	83×2	84×2
# 2	80×3	81×3	82×3	83×3	84×3
# 3	80×4	81×4	82×4	83×4	84×4
# 4	80×5	81×5	82×5	83×5	84×5
# 5	80×6	81×6	82×6	83×6	84×6
# 6	80×7	81×7	82×7	83×7	84×7
# 7	80×8	81×8	82×8	83×8	84×8
# 8	80×9	81×9	82×9	83×9	84×9
# 9	85×2	86×2	87×2	88×2	89×2
# 10	85×3	86×3	87×3	88×3	89×3

Grade 4 Chapter 2 Lesson 10

Multiply numbers in column

6	1) $9 \times 7 = 63$
87	2) Put 3 below 9
$\times\underline{9}$	3) Put 6 above 8
783	4) $9 \times 8 = 72$
	5) $72 + 6 = 78$
	6) Put 78 ahead 3

Multiply numbers

	a	b	c	d	e
# 11	85×4	86×4	87×4	88×4	89×4
# 12	85×5	86×5	87×5	88×5	89×5
# 13	85×6	86×6	87×6	88×6	89×6
# 14	85×7	86×7	87×7	88×7	89×7
# 15	85×8	86×8	87×8	88×8	89×8
# 16	85×9	86×9	87×9	88×9	89×9

Grade 4 Chapter 2 Lesson 11
Multiplication from 90 to 99
Multiply numbers in row

1) 96×7
2) $96 = 90 + 6$
3) $90 \times 7 = 630$
4) $6 \times 7 = 42$
5) $630 + 42 = 672$
6) $96\times 7 = 672$

Multiply numbers

	a	b	c	d	e
# 1	90×2	91×2	92×2	93×2	94×2
# 2	90×3	91×3	92×3	93×3	94×3
# 3	90×4	91×4	92×4	93×4	94×4
# 4	90×5	91×5	92×5	93×5	94×5
# 5	90×6	91×6	92×6	93×6	94×6
# 6	90×7	91×7	92×7	93×7	94×7
# 7	90×8	91×8	92×8	93×8	94×8
# 8	90×9	91×9	92×9	93×9	94×9
# 9	95×2	96×2	97×2	98×2	99×2
# 10	95×3	96×3	97×3	98×3	99×3

Grade 4 Chapter 2 Lesson 11
Multiply numbers in column

$$\begin{array}{r} {}^{4}\\ 96 \\ \times \underline{7} \\ 672 \end{array}$$

1) 7 × 6 = 42
2) Put 2 below 7
3) Put 4 above 9
4) 7 × 9 = 63
5) 63 + 4 = 67
6) Put 67 ahead 2

Multiply numbers

	a	b	c	d	e
# 11	95 × 4	96 × 4	97 × 4	98 × 4	99 × 4
# 12	95 × 5	96 × 5	97 × 5	98 × 5	99 × 5
# 13	95 × 6	96 × 6	97 × 6	98 × 6	99 × 6
# 14	95 × 7	96 × 7	97 × 7	98 × 7	99 × 7
# 15	95 × 8	96 × 8	97 × 8	98 × 8	99 × 8
# 16	95 × 9	96 × 9	97 × 9	98 × 9	99 × 9

Grade 4 Chapter 2 Lesson 12
Multiplication and Addition numbers
To multiply and add numbers always first multiply than add

$$9 \times 6 + 9 = 54 + 9 = 63$$

Evaluate

	a	b	c	d
# 1	$2 \times 1 + 0$	$3 \times 1 + 0$	$4 \times 1 + 0$	$5 \times 1 + 0$
# 2	$2 \times 2 + 1$	$3 \times 2 + 1$	$4 \times 2 + 1$	$5 \times 2 + 1$
# 3	$2 \times 3 + 2$	$3 \times 3 + 2$	$4 \times 3 + 2$	$5 \times 3 + 2$
# 4	$2 \times 4 + 3$	$3 \times 4 + 3$	$4 \times 4 + 3$	$5 \times 4 + 3$
# 5	$2 \times 5 + 4$	$3 \times 5 + 4$	$4 \times 5 + 4$	$5 \times 5 + 4$
# 6	$2 \times 6 + 5$	$3 \times 6 + 5$	$4 \times 6 + 5$	$5 \times 6 + 5$
# 7	$2 \times 7 + 6$	$3 \times 7 + 6$	$4 \times 7 + 6$	$5 \times 7 + 6$
# 8	$2 \times 8 + 7$	$3 \times 8 + 7$	$4 \times 8 + 7$	$5 \times 8 + 7$
# 9	$2 \times 9 + 8$	$3 \times 9 + 8$	$4 \times 9 + 8$	$5 \times 9 + 8$
# 10	$2 \times 10 + 9$	$3 \times 10 + 9$	$4 \times 10 + 9$	$5 \times 10 + 9$

Multiply and add numbers

$$9 \overset{2}{+} 9 \overset{1}{\times} 9 = 9 + 81 = 90$$

Evaluate

	a	b	c	d
# 11	$0 + 6 \times 1$	$0 + 7 \times 1$	$0 + 8 \times 1$	$0 + 9 \times 1$
# 12	$1 + 6 \times 2$	$1 + 7 \times 2$	$1 + 8 \times 2$	$1 + 9 \times 2$
# 13	$2 + 6 \times 3$	$2 + 7 \times 3$	$2 + 8 \times 3$	$2 + 9 \times 3$
# 14	$3 + 6 \times 4$	$3 + 7 \times 4$	$3 + 8 \times 4$	$3 + 9 \times 4$
# 15	$4 + 6 \times 5$	$4 + 7 \times 5$	$4 + 8 \times 5$	$4 + 9 \times 5$
# 16	$5 + 6 \times 6$	$5 + 7 \times 6$	$5 + 8 \times 6$	$5 + 9 \times 6$
# 17	$6 + 6 \times 7$	$6 + 7 \times 7$	$6 + 8 \times 7$	$6 + 9 \times 7$
# 18	$7 + 6 \times 8$	$7 + 7 \times 8$	$7 + 8 \times 8$	$7 + 9 \times 8$
# 19	$8 + 6 \times 9$	$8 + 7 \times 9$	$8 + 8 \times 9$	$8 + 9 \times 9$
# 20	$9 + 6 \times 10$	$9 + 7 \times 10$	$9 + 8 \times 10$	$9 + 9 \times 10$

Grade 4 Chapter 2 Lesson 12
Multiplication and Subtraction numbers
To multiply and subtraction numbers always first multiply than subtract

$5 \times 10 - 9 = 50 - 9 = 41$

Evaluate

	a	b	c	d
# 21	2 × 1 - 0	3 × 1 - 0	4 × 1 - 0	5 × 1 - 0
# 22	2 × 2 - 1	3 × 2 - 1	4 × 2 - 1	5 × 2 - 1
# 23	2 × 3 - 2	3 × 3 - 2	4 × 3 - 2	5 × 3 - 2
# 24	2 × 4 - 3	3 × 4 - 3	4 × 4 - 3	5 × 4 - 3
# 25	2 × 5 - 4	3 × 5 - 4	4 × 5 - 4	5 × 5 - 4
# 26	2 × 6 - 5	3 × 6 - 5	4 × 6 - 5	5 × 6 - 5
# 27	2 × 7 - 6	3 × 7 - 6	4 × 7 - 6	5 × 7 - 6
# 28	2 × 8 - 7	3 × 8 - 7	4 × 8 - 7	5 × 8 - 7
# 29	2 × 9 - 8	3 × 9 - 8	4 × 9 - 8	5 × 9 - 8
# 30	2 × 10 - 9	3 × 10 - 9	4 × 10 - 9	5× 10 - 9

Multiply and Subtract numbers

$100 - \overset{2}{9 \times} \overset{1}{10} = 100 - 90 = 10$

Evaluate

	a	b	c	d
# 31	10 - 6× 1	10 - 7× 1	10 - 8× 1	10 - 9 × 1
# 32	20 - 6× 2	20 - 7× 2	20 - 8× 2	20 - 9 × 2
# 33	30 - 6× 3	30 - 7× 3	30 - 8 × 3	30 - 9 × 3
# 34	40 - 6× 4	40 - 7× 4	40 - 8 × 4	40 - 9 × 4
# 35	50 - 6× 5	50 - 7× 5	50 - 8 × 5	50 - 9 × 5
# 36	60 - 6× 6	60 - 7× 6	60 - 8 × 6	60 - 9 × 6
# 37	70 - 6× 7	70 - 7× 7	70 - 8 × 7	70 - 9 × 7
# 38	80 - 6× 8	80 - 7× 8	80 - 8 × 8	80 - 9 × 8
# 39	90 - 6× 9	90 - 7× 9	90 - 8 × 9	90 - 9 × 9
# 40	100 -6× 10	100 -7× 10	100 - 8× 10	100 - 9× 10

Grade 4 Chapter 2 Lesson 13

Summary 1

Three-digit numbers

Three-digit numbers content three digits.
This is numbers 100, 101, … 999
Least three-digit number: 100.
Greatest three-digit number: 999
Digits for three-digit numbers from right to the left:
Ones, tens, and hundreds $468 = 400 + 60 + 8$

One-digit numbers less than two or three-digit numbers
Three-digit numbers greater than one or two-digit numbers

To compare numbers, compare first hundred, after that tens, then ones

$423 < 456$ $\qquad$ $789 > 786$

Two numbers we can add in any order. The sum will be the same.

$200 + 300 = 500$ $\qquad$ $300+200=500$

The sum of a number and zero is the same number

$500+0 = 500$ $\qquad$ $0+500=500$

Difference any number and zero is the same number:

$500 - 0 = 500$

Difference two the same number is zero: $500 - 500 = 0$

Adding three-digit numbers: Add the ones. Add the tens. Add the hundreds.

Subtracting three-digit numbers:
Subtract the ones. Subtract the tens. Subtract the hundreds.

		11	71414
246	678	154	85 4
+ 351	- 432	+789	-28 6
597	246	943	56 8

Grade 4 Chapter 2 Lesson 13
Summary 2
Multiplication numbers

Third operation with numbers is multiplication $\times$ or $\cdot$
Multiplication is the operation of repeated addition
$2\times 3 = 2 + 2 + 2 = 6$ or $2 \cdot 3 = 2 + 2 + 2 = 6$

First number in multiplication called the factor	**2**
Second number in multiplication called the factor	**3**
Result of multiplication called the product	**6**

Properties of multiplication
Two numbers we can multiply in any order. The product will the same
$2 \times 3 = 2 + 2 + 2 = 6$ $3 \times 2 = 3 + 3 = 6$
The product of a number and zero is zero
$5 \times 0 = 0$ $0 \times 5 = 0$
The product of a number and one is the same number
$5 \times 1 = 5$ $1 \times 5 = 5$

Division two-digit and one-digit numbers
Fourth operation with numbers is division $\div$ or :
$6 \div 2 = 3$ $6 : 3 = 2$
Division means separated number by some equal groups

First number in division is called dividend	**6**
Second number in division is called divisor	**2**
Result of division is called quotient	**3**

Properties of division

Any number divided by one is the same number		**$5 \div 1 = 5$**
Any number divided by the itself is one		**$5 \div 5 = 1$**
Zero divided by any number is zero		**$0 \div 5 = 0$**
Division by zero is not allowed	**$5 \div 0$**	**No number**

Grade 4 Chapter 2 Lesson 13
Review 2 A
Multiply the numbers in column

# 1 $\begin{array}{r} 16 \\ \times 9 \\ \hline \end{array}$	# 2 $\begin{array}{r} 18 \\ \times 7 \\ \hline \end{array}$	# 3 $\begin{array}{r} 26 \\ \times 7 \\ \hline \end{array}$	# 4 $\begin{array}{r} 28 \\ \times 6 \\ \hline \end{array}$
# 5 $\begin{array}{r} 36 \\ \times 7 \\ \hline \end{array}$	# 6 $\begin{array}{r} 38 \\ \times 9 \\ \hline \end{array}$	# 7 $\begin{array}{r} 46 \\ \times 8 \\ \hline \end{array}$	# 8 $\begin{array}{r} 48 \\ \times 7 \\ \hline \end{array}$
# 9 $\begin{array}{r} 56 \\ \times 7 \\ \hline \end{array}$	# 10 $\begin{array}{r} 58 \\ \times 5 \\ \hline \end{array}$	# 11 $\begin{array}{r} 65 \\ \times 9 \\ \hline \end{array}$	# 12 $\begin{array}{r} 68 \\ \times 9 \\ \hline \end{array}$
# 13 $\begin{array}{r} 76 \\ \times 8 \\ \hline \end{array}$	# 14 $\begin{array}{r} 78 \\ \times 9 \\ \hline \end{array}$	# 15 $\begin{array}{r} 86 \\ \times 7 \\ \hline \end{array}$	# 16 $\begin{array}{r} 89 \\ \times 7 \\ \hline \end{array}$
# 17 $\begin{array}{r} 92 \\ \times 6 \\ \hline \end{array}$	# 18 $\begin{array}{r} 95 \\ \times 7 \\ \hline \end{array}$	# 19 $\begin{array}{r} 96 \\ \times 7 \\ \hline \end{array}$	# 20 $\begin{array}{r} 98 \\ \times 6 \\ \hline \end{array}$

Grade 4 Chapter 2 Lesson 13

Review 2 B

Multiply the numbers in column

# 1 17×9	# 2 18×6	# 3 26×9	# 4 28×7
# 5 36×5	# 6 38×7	# 7 46×8	# 8 48×7
# 9 56×8	# 10 58×6	# 11 67×9	# 12 68×7
# 13 76×9	# 14 78×6	# 15 86×5	# 16 87×9
# 17 92×8	# 18 94×7	# 19 96×8	# 20 98×5

Grade 4 Chapter 2 Lesson 14
Multiplication numbers from 10 to 99

Test 2 A

Multiply the numbers in column

1 $\begin{array}{r} 15 \\ \times\ 9 \\ \hline \end{array}$ # 2 $\begin{array}{r} 16 \\ \times\ 7 \\ \hline \end{array}$ # 3 $\begin{array}{r} 25 \\ \times\ 7 \\ \hline \end{array}$ # 4 $\begin{array}{r} 27 \\ \times\ 6 \\ \hline \end{array}$

5 $\begin{array}{r} 34 \\ \times\ 7 \\ \hline \end{array}$ # 6 $\begin{array}{r} 36 \\ \times\ 9 \\ \hline \end{array}$ # 7 $\begin{array}{r} 45 \\ \times\ 8 \\ \hline \end{array}$ # 8 $\begin{array}{r} 46 \\ \times\ 7 \\ \hline \end{array}$

9 $\begin{array}{r} 54 \\ \times\ 7 \\ \hline \end{array}$ # 10 $\begin{array}{r} 56 \\ \times\ 5 \\ \hline \end{array}$ # 11 $\begin{array}{r} 64 \\ \times\ 9 \\ \hline \end{array}$ # 12 $\begin{array}{r} 66 \\ \times\ 9 \\ \hline \end{array}$

13 $\begin{array}{r} 74 \\ \times\ 8 \\ \hline \end{array}$ # 14 $\begin{array}{r} 76 \\ \times\ 9 \\ \hline \end{array}$ # 15 $\begin{array}{r} 84 \\ \times\ 7 \\ \hline \end{array}$ # 16 $\begin{array}{r} 85 \\ \times\ 7 \\ \hline \end{array}$

17 $\begin{array}{r} 94 \\ \times\ 6 \\ \hline \end{array}$ # 18 $\begin{array}{r} 96 \\ \times\ 7 \\ \hline \end{array}$ # 19 $\begin{array}{r} 94 \\ \times\ 7 \\ \hline \end{array}$ # 20 $\begin{array}{r} 96 \\ \times\ 6 \\ \hline \end{array}$

Grade 4 Chapter 2 Lesson 14
Multiplication numbers from 10 to 99

Test 2 B

Multiply the numbers in column

# 1 $\begin{array}{r} 14 \\ \times 9 \\ \hline \end{array}$	# 2 $\begin{array}{r} 19 \\ \times 6 \\ \hline \end{array}$	# 3 $\begin{array}{r} 25 \\ \times 9 \\ \hline \end{array}$	# 4 $\begin{array}{r} 26 \\ \times 7 \\ \hline \end{array}$
# 5 $\begin{array}{r} 34 \\ \times 5 \\ \hline \end{array}$	# 6 $\begin{array}{r} 36 \\ \times 7 \\ \hline \end{array}$	# 7 $\begin{array}{r} 48 \\ \times 8 \\ \hline \end{array}$	# 8 $\begin{array}{r} 46 \\ \times 7 \\ \hline \end{array}$
# 9 $\begin{array}{r} 54 \\ \times 8 \\ \hline \end{array}$	# 10 $\begin{array}{r} 56 \\ \times 6 \\ \hline \end{array}$	# 11 $\begin{array}{r} 65 \\ \times 9 \\ \hline \end{array}$	# 12 $\begin{array}{r} 69 \\ \times 7 \\ \hline \end{array}$
# 13 $\begin{array}{r} 74 \\ \times 9 \\ \hline \end{array}$	# 14 $\begin{array}{r} 76 \\ \times 6 \\ \hline \end{array}$	# 15 $\begin{array}{r} 84 \\ \times 5 \\ \hline \end{array}$	# 16 $\begin{array}{r} 86 \\ \times 9 \\ \hline \end{array}$
# 17 $\begin{array}{r} 94 \\ \times 8 \\ \hline \end{array}$	# 18 $\begin{array}{r} 96 \\ \times 7 \\ \hline \end{array}$	# 19 $\begin{array}{r} 94 \\ \times 8 \\ \hline \end{array}$	# 20 $\begin{array}{r} 96 \\ \times 5 \\ \hline \end{array}$

Grade 4 Answers Chapter 2

Lesson 2	a	b	c	d		a	b	c	d
# 1	2,	3,	4,	5	# 11	6,	7,	8,	9
# 2	4,	6,	8,	10	# 12	12,	14,	16,	18
# 3	6,	9,	12,	15	# 13	18,	21,	24,	27
# 4	8,	12,	16,	20	# 14	24,	28,	32,	36
# 5	10,	15,	20,	25	# 15	30,	35,	40,	45
# 6	12,	18,	24,	30	# 16	36,	42,	48,	54
# 7	14,	21,	28,	35	# 17	42,	49,	56,	63
# 8	16,	24,	32,	40	# 18	48,	56,	64,	72
# 9	18,	27,	36,	45	# 19	54,	63,	72,	81
# 10	20,	30,	40,	50	# 20	60,	70,	80,	90

Lesson 3	a	b	c	d	e		a	b	c	d	e
# 1	20,	22,	24,	26,	28	# 9	30,	32,	34,	36,	38
# 2	30,	33,	36,	39,	42	# 10	45,	48,	51,	54,	57
# 3	40,	44,	48,	52,	56	# 11	60,	64,	68,	72,	76
# 4	50,	55,	60,	65,	70	# 12	75,	80,	85,	90,	95
# 5	60,	66,	72,	78,	84	# 13	90,	96,	102,	108,	114
# 6	70,	77,	84,	91,	98	# 14	105,	112,	119,	126,	133
# 7	80,	88,	96,	104,	112	# 15	120,	128,	136,	144,	152
# 8	90,	99,	108,	117,	126	# 16	135,	144,	153,	162,	171

Lesson 4	a	b	c	d	e		a	b	c	d	e
# 1	40,	42,	44,	46,	48	# 9	50,	52,	54,	56,	58
# 2	60,	63,	66,	69,	72	# 10	75,	78,	81,	84,	87
# 3	80,	84,	88,	92,	96	# 11	100,	104,	108,	112,	116
# 4	100,	105,	110,	115,	120	# 12	125,	130,	135,	140,	145
# 5	120,	126,	132,	138,	144	# 13	150,	156,	162,	168,	174
# 6	140,	147,	154,	161,	168	# 14	175,	182,	189,	196,	203
# 7	160,	168,	176,	184,	192	# 15	200,	208,	216,	224,	232
# 8	180,	189,	198,	207,	216	# 16	225,	234,	243,	252,	261

Lesson 5	a	b	c	d	e		a	b	c	d	e
# 1	60,	62,	64,	66,	68	# 9	70,	72,	74,	76,	78
# 2	90,	93,	96,	99,	102	# 10	105,	108,	111,	114,	117
# 3	120,	124,	128,	132,	136	# 11	140,	144,	148,	152,	156
# 4	150,	155,	160,	165,	170	# 12	175,	180,	185,	190,	195
# 5	180,	186,	192,	198,	204	# 13	210,	216,	222,	228,	234
# 6	210,	217,	224,	231,	238	# 14	245,	252,	259,	266,	273
# 7	240,	248,	256,	264,	272	# 15	280,	288,	296,	304,	312
# 8	270,	279,	288,	297,	306	# 16	315,	324,	333,	342,	351

Grade 4 Answers Chapter 2

Lesson 6	a b c d e		a b c d e
# 1	80, 82, 84, 86, 88	# 9	90, 92, 94, 96, 98
# 2	120, 123, 126, 129, 132	# 10	135, 138, 141, 144, 147
# 3	160, 164, 168, 172, 176	# 11	180, 184, 188, 192, 196
# 4	200, 205, 210, 215, 220	# 12	225, 230, 235, 240, 245
# 5	240, 246, 252, 258, 264	# 13	270, 276, 282, 288, 294
# 6	280, 287, 294, 301, 308	# 14	315, 322, 329, 336, 343
# 7	320, 328, 336, 344, 352	# 15	360, 368, 376, 384, 392
# 8	360, 369, 378, 387, 396	# 16	405, 414, 423, 432, 441

Lesson 7	a b c d e		a b c d e
# 1	100, 102, 104, 106, 108	# 9	110, 112, 114, 116, 118
# 2	150, 153, 156, 159, 162	# 10	165, 168, 171, 174, 177
# 3	200, 204, 208, 212, 216	# 11	220, 224, 228, 232, 236
# 4	250, 255, 260, 265, 270	# 12	275, 280, 285, 290, 295
# 5	300, 306, 312, 318, 324	# 13	330, 336, 342, 348, 354
# 6	350, 357, 364, 371, 378	# 14	385, 392, 399, 406, 413
# 7	400, 408, 416, 424, 432	# 15	440, 448, 456, 464, 472
# 8	450, 459, 468, 477, 486	# 16	495, 504, 513, 522, 531

Lesson 8	a b c d e		a b c d e
# 1	120, 122, 124, 126, 128	# 9	130, 132, 134, 136, 138
# 2	180, 183, 186, 189, 192	# 10	195, 198, 201, 204, 207
# 3	240, 244, 248, 252, 256	# 11	260, 264, 268, 272, 276
# 4	300, 305, 310, 315, 320	# 12	325, 330, 335, 340, 345
# 5	360, 366, 372, 378, 384	# 13	390, 396, 402, 408, 414
# 6	420, 427, 434, 441, 448	# 14	455, 462, 469, 476, 483
# 7	480, 488, 496, 504, 512	# 15	520, 528, 536, 544, 552
# 8	540, 549, 558, 567, 576	# 16	585, 594, 603, 612, 621

Lesson 9	a b c d e		a b c d e
# 1	140, 142, 144, 146, 148	# 9	150, 152, 154, 156, 158
# 2	210, 213, 216, 219, 222	# 10	225, 228, 231, 234, 237
# 3	280, 284, 288, 292, 296	# 11	300, 304, 308, 312, 316
# 4	350, 355, 360, 365, 370	# 12	375, 380, 385, 390, 395
# 5	420, 426, 432, 438, 444	# 13	450, 456, 462, 468, 474
# 6	490, 497, 504, 511, 518	# 14	525, 532, 539, 546, 553
# 7	560, 568, 576, 584, 592	# 15	600, 608, 616, 624, 632
# 8	630, 639, 648, 657, 666	# 16	675, 684, 693, 702, 711

Grade 4 Answers Chapter 2

Lesson 10	a b c d e		a b c d e
# 1	160, 162, 164, 166, 168	# 9	170, 172, 174, 176, 178
# 2	240, 243, 246, 249, 252	# 10	255, 258, 261, 264, 267
# 3	320, 324, 328, 332, 336	# 11	340, 344, 348, 352, 356
# 4	400, 405, 410, 415, 420	# 12	425, 430, 435, 440, 445
# 5	480, 486, 492, 498, 504	# 13	510, 516, 522, 528, 534
# 6	560, 567, 574, 581, 588	# 14	595, 602, 609, 616, 623
# 7	640, 648, 656, 664, 672	# 15	680, 688, 696, 704, 712
# 8	720, 729, 738, 747, 756	# 16	765, 774, 783, 792, 801

Lesson 11	a b c d e		a b c d e
# 1	180, 182, 184, 186, 188	# 9	190, 192, 194, 196, 198
# 2	270, 273, 276, 279, 282	# 10	285, 288, 291, 294, 297
# 3	360, 364, 368, 372, 376	# 11	380, 384, 388, 392, 396
# 4	450, 455, 460, 465, 470	# 12	475, 480, 485, 490, 495
# 5	540, 546, 552, 558, 564	# 13	570, 576, 582, 588, 594
# 6	630, 637, 644, 651, 658	# 14	665, 672, 679, 686, 693
# 7	720, 728, 736, 744, 752	# 15	760, 768, 776, 784, 792
# 8	810, 819, 828, 837, 846	# 16	855, 864, 873, 882, 891

Lesson 12	a	b	c	d		a	b	c	d
# 1	2	3	4	5	# 21	2	3	4	5
# 2	5	7	9	11	# 22	3	5	7	9
# 3	8	11	14	17	# 23	4	7	10	13
# 4	11	15	19	23	# 24	5	9	13	17
# 5	14	19	24	29	# 25	6	11	16	21
# 6	17	23	29	35	# 26	7	13	19	25
# 7	20	27	34	41	# 27	8	15	22	29
# 8	23	31	39	47	# 28	9	17	25	33
# 9	26	35	44	53	# 29	10	19	28	37
# 10	29	39	49	59	# 30	11	21	31	41
# 11	6	7	8	9	# 31	4	3	2	1
# 12	13	15	17	19	# 32	8	6	4	2
# 13	20	23	26	29	# 33	12	9	6	3
# 14	27	31	35	39	# 34	16	12	8	4
# 15	34	39	44	49	# 35	20	15	10	5
# 16	41	47	53	59	# 36	24	18	12	6
# 17	48	55	62	69	# 37	28	21	14	7
# 18	55	63	71	79	# 38	32	24	16	8
# 19	62	71	80	89	# 39	36	27	18	9
# 20	69	79	89	99	# 40	40	30	20	10

Grade 4 Answers Chapter 2

Lesson 13

Review 2 A

# 1	144	# 6	342	# 11	585	# 16	623
# 2	126	# 7	368	# 12	612	# 17	552
# 3	182	# 8	336	# 13	608	# 18	665
# 4	168	# 9	392	# 14	702	# 19	672
# 5	252	# 10	290	# 15	602	# 20	588

Lesson 13

Review 2 B

# 1	153	# 6	266	# 11	603	# 16	783
# 2	108	# 7	368	# 12	476	# 17	736
# 3	234	# 8	336	# 13	684	# 18	658
# 4	196	# 9	448	# 14	468	# 19	768
# 5	180	# 10	348	# 15	430	# 20	490

Lesson 14

Test 2 A

# 1	135	# 2	112	# 3	175	# 4	162	# 5	238
# 6	252	# 7	360	# 8	322	# 9	378	# 10	280
# 11	576	# 12	594	# 13	592	# 14	684	# 15	588
# 16	595	# 17	564	# 18	672	# 19	658	# 20	576

Test 2 B

# 1	126	# 2	114	# 3	225	# 4	182	# 5	170
# 6	252	# 7	384	# 8	322	# 9	432	# 10	336
# 11	585	# 12	483	# 13	666	# 14	456	# 15	420
# 16	774	# 17	752	# 18	672	# 19	752	# 20	480

Grade 4

Chapter 3

Division

Numbers

Grade 4 Chapter 3 Contents
Division Numbers. Lesson 1 – 18
Basic Problems

Work with lesson 1, read rules and examples, 2) Solve the problems from contents, 3) Check answers in the end of chapter 1, 4) Work with the next lessons 2, 3, 4, 5, …, 5) Solve Test 1 A, 6) Grading Scale: A =10 points, B=9points, C=8 points, D=7 points, F=0-6 points, 7) If you solve 10 problems correctly, you get an A, 8) If have an A, you can move on to next chapter, 9) If your grade is lower than an A, work with the mistakes, then solve Test 1B, 10) When you finish all of the chapters, solve Final Exam in last chapter.

Summary 4 Chapter 3 Lesson 1
Division numbers 1

Fourth operation with numbers is division ÷ or :
Division means repeated subtraction. Result of division means how many times we need to subtract the same number to get zero

$6 - 2 - 2 - 2 = 0$	$6 \div 2 = 3$	$6 - 3 - 3 = 0$	$6 : 3 = 2$
1 2 3		1 2	

Repeat table multiplication on page 32

$2 \times 3 = 6 \quad 6 \div 2 = 3 \quad 6 \div 3 = 2$

Divide the numbers using table multiplication

	a	b	c	d
# 1	$2 \div 2$	$12 \div 2$	$4 \div 4$	$24 \div 4$
# 2	$4 \div 2$	$14 \div 2$	$8 \div 4$	$28 \div 4$
# 3	$6 \div 2$	$16 \div 2$	$12 \div 4$	$32 \div 4$
# 4	$8 \div 2$	$18 \div 2$	$16 \div 4$	$36 \div 4$
# 5	$10 \div 2$	$20 \div 2$	$20 \div 4$	$40 \div 4$
# 6	$3 \div 3$	$18 \div 3$	$5 \div 5$	$30 \div 5$
# 7	$6 \div 3$	$21 \div 3$	$10 \div 5$	$35 \div 5$
# 8	$9 \div 3$	$24 \div 3$	$15 \div 5$	$40 \div 5$
# 9	$12 \div 3$	$27 \div 3$	$20 \div 5$	$45 \div 5$
# 10	$15 \div 3$	$30 \div 3$	$25 \div 5$	$50 \div 5$

Grade 4 Chapter 3 Lesson 2
Properties of division 2

Division also means separated number by some equal groups

$6 \div 2 = 3$ $2 \times 3 = 6$

First number in division called the dividend 6

Second number in division called the divisor 2

Result of division called the quotient 3

Any number divided by one is the same number $5 \div 1 = 5$

Any number divided by the itself is one $5 \div 5 = 1$

Zero divided by any number is zero $0 \div 5 = 0$

Division by zero is not allowed $5 \div 0$ No number

To check multiplication use division

$2 \times 3 = 6$ Check $6 \div 2 = 3$ or $6 \div 3 = 2$

To check division use multiplication and division

$6 \div 2 = 3$ Check $2 \times 3 = 6$ or $6 \div 3 = 2$

Division without remainder R = 0

Dividend 35, Divisor 7, Quotient 5

$$\begin{array}{r} 5 \\ 7\overline{)35} \\ -35 \\ \hline 0 \end{array}$$

R = 0

Division with remainder R > 0

Dividend 36, Divisor 7, Quotient 5

$$\begin{array}{r} 5 \\ 7\overline{)36} \\ -35 \\ \hline 1 \end{array}$$

R = 1

Grade 4 Chapter 3 Lesson 2
Divide numbers by using Table of multiplication

	a	b	c	d
# 1	0 ÷ 6	7 ÷ 7	72 ÷ 8	0 ÷ 9
# 2	12 ÷6	21 ÷ 7	64 ÷ 8	63 ÷ 9
# 3	30 ÷ 6	28 ÷ 7	48 ÷ 8	45 ÷ 9
# 4	42 ÷ 6	42 ÷ 7	32 ÷ 8	27 ÷ 9
# 5	48 ÷ 6	63 ÷ 7	16 ÷ 8	9 ÷ 9
# 6	54 ÷ 6	56 ÷ 7	8 ÷ 8	18 ÷ 9
# 7	36 ÷ 6	49 ÷ 7	24 ÷ 8	36 ÷ 9
# 8	24 ÷ 6	35 ÷ 7	40 ÷ 8	54 ÷ 9
# 9	18 ÷ 6	14 ÷ 7	56 ÷ 8	72 ÷ 9
# 10	6 ÷ 6	0 ÷ 7	0 ÷ 8	81 ÷ 9

Complete the Table of Division by using Table of Multiplication

	×	2		4		6		8	
#1	1								9
#2									18
#3	3							24	
#4								32	
#5	5						35		
#6						36			
#7	7				35				
#8				32					
#9	9		27						
#10		20							

Grade 4 Chapter 3 Lesson 3
Division number with quotient from 10 to 19
Divide numbers in column 3

Divide numbers

	a	b	c	d
# 1	20 ÷ 2	33 ÷ 3	76 ÷ 4	90 ÷ 5
# 2	24 ÷ 2	39 ÷ 3	68 ÷ 4	80 ÷ 5
# 3	28 ÷ 2	45 ÷ 3	60 ÷ 4	70 ÷ 5
# 4	32 ÷ 2	51 ÷ 3	52 ÷ 4	60 ÷ 5
# 5	36 ÷ 2	57 ÷ 3	44 ÷ 4	50 ÷ 5
# 6	22 ÷ 2	30 ÷ 3	72 ÷ 4	95 ÷ 5
# 7	26 ÷ 2	36 ÷ 3	64 ÷ 4	85 ÷ 5
# 8	30 ÷ 2	42 ÷ 3	56 ÷ 4	75 ÷ 5
# 9	34 ÷ 2	48 ÷ 3	48 ÷ 4	65 ÷ 5
# 10	38 ÷ 2	54 ÷ 3	40 ÷ 4	55 ÷ 5

Divide numbers in column

$9 \times 1 = 9 \quad 9 \times 2 = 18$

$$\begin{array}{r} 19 \\ 9\,\overline{)\,171} \\ \underline{-9} \\ 81 \\ \underline{-81} \\ 0 \end{array}$$

1) $9 \times 1 = 9$
2) $17 - 9 = 8$
3) Bring down the 1
4) $9 \times 9 = 81$
5) $81 - 81 = 0$
6) $R = 0$

Check

$$\begin{array}{r} 8 \\ 19 \\ \underline{\times 9} \\ 171 \end{array}$$

Divide numbers

	a	b	c	d
# 11	60 ÷ 6	77 ÷ 7	152 ÷ 8	162 ÷ 9
# 12	72 ÷ 6	91 ÷ 7	136 ÷ 8	144 ÷ 9
# 13	84 ÷ 6	105 ÷ 7	120 ÷ 8	126 ÷ 9
# 14	96 ÷ 6	119 ÷ 7	104 ÷ 8	108 ÷ 9
# 15	108 ÷ 6	133 ÷ 7	88 ÷ 8	90 ÷ 9
# 16	66 ÷ 6	70 ÷ 7	144 ÷ 8	171 ÷ 9
# 17	78 ÷ 6	84 ÷ 7	128 ÷ 8	153 ÷ 9
# 18	90 ÷ 6	98 ÷ 7	112 ÷ 8	135 ÷ 9
# 19	102 ÷ 6	112 ÷ 7	96 ÷ 8	117 ÷ 9
# 20	114 ÷ 6	126 ÷ 7	80 ÷ 8	99 ÷ 9

Grade 4 Chapter 3 Lesson 4
Division numbers with quotient from 20 to 29
Divide numbers in column 4

Divide numbers

	a	b	c	d
# 1	$40 \div 2$	$63 \div 3$	$116 \div 4$	$140 \div 5$
# 2	$44 \div 2$	$69 \div 3$	$108 \div 4$	$130 \div 5$
# 3	$48 \div 2$	$75 \div 3$	$100 \div 4$	$120 \div 5$
# 4	$52 \div 2$	$81 \div 3$	$92 \div 4$	$110 \div 5$
# 5	$56 \div 2$	$87 \div 3$	$84 \div 4$	$100 \div 5$
# 6	$42 \div 2$	$60 \div 3$	$112 \div 4$	$145 \div 5$
# 7	$46 \div 2$	$66 \div 3$	$104 \div 4$	$135 \div 5$
# 8	$50 \div 2$	$72 \div 3$	$96 \div 4$	$125 \div 5$
# 9	$54 \div 2$	$78 \div 3$	$88 \div 4$	$115 \div 5$
# 10	$58 \div 2$	$84 \div 3$	$80 \div 4$	$105 \div 5$

Divide numbers in column

$9 \times 1 = 9$ $9 \times 2 = 18$ $9 \times 3 = 27$

$$\begin{array}{r} 29 \\ 9\overline{)261} \\ \underline{-18} \\ 81 \\ \underline{-81} \\ 0 \end{array}$$

1) $9 \times 2 = 18$
2) $26 - 18 = 8$
3) Bring down the 1
4) $9 \times 9 = 81$
5) $81 - 81 = 0$
6) $R = 0$

or

$$\begin{array}{rl} 261 & \underline{|\,9} \\ -\underline{18} & 29 \\ 81 & \\ \underline{-81} & \\ 0 & \end{array}$$

Divide numbers

	a	b	c	d
# 11	$120 \div 6$	$147 \div 7$	$232 \div 8$	$252 \div 9$
# 12	$132 \div 6$	$161 \div 7$	$216 \div 8$	$234 \div 9$
# 13	$144 \div 6$	$175 \div 7$	$200 \div 8$	$216 \div 9$
# 14	$156 \div 6$	$189 \div 7$	$184 \div 8$	$198 \div 9$
# 15	$168 \div 6$	$203 \div 7$	$168 \div 8$	$180 \div 9$
# 16	$126 \div 6$	$140 \div 7$	$224 \div 8$	$261 \div 9$
# 17	$138 \div 6$	$154 \div 7$	$208 \div 8$	$243 \div 9$
# 18	$150 \div 6$	$168 \div 7$	$192 \div 8$	$225 \div 9$
# 19	$162 \div 6$	$182 \div 7$	$176 \div 8$	$207 \div 9$
# 20	$174 \div 6$	$196 \div 7$	$160 \div 8$	$189 \div 9$

Grade 4 Chapter 3 Lesson 5

Division numbers with quotient from 30 to 39

Divide numbers in column 5

Divide numbers

	a	b	c	d
# 1	60 ÷ 2	93 ÷ 3	156 ÷ 4	190 ÷ 5
# 2	64 ÷ 2	99 ÷ 3	148 ÷ 4	180 ÷ 5
# 3	68 ÷ 2	105 ÷ 3	140 ÷ 4	170 ÷ 5
# 4	72 ÷ 2	111 ÷ 3	132 ÷ 4	160 ÷ 5
# 5	76 ÷ 2	117 ÷ 3	124 ÷ 4	150 ÷ 5
# 6	62 ÷ 2	90 ÷ 3	152 ÷ 4	195 ÷ 5
# 7	66 ÷ 2	96 ÷ 3	144 ÷ 4	185 ÷ 5
# 8	70 ÷ 2	102 ÷ 3	136 ÷ 4	175 ÷ 5
# 9	74 ÷ 2	108 ÷ 3	128 ÷ 4	165 ÷ 5
# 10	78 ÷ 2	114 ÷ 3	120 ÷ 4	155 ÷ 5

Divide numbers in column

39	1) 9× 3 = 27
9 | 351	2) 35 – 27 = 8
- 27	3) Bring down the 1
81	4) 9 × 9 = 81
- 81	5) 81 – 81 = 0
0	6) R = 0

Divide numbers

	a	b	c	d
# 11	180 ÷ 6	217 ÷ 7	312 ÷ 8	342 ÷ 9
# 12	192 ÷ 6	231 ÷ 7	296 ÷ 8	324 ÷ 9
# 13	204 ÷ 6	245 ÷ 7	280 ÷ 8	306 ÷ 9
# 14	216 ÷ 6	259 ÷ 7	264 ÷ 8	288 ÷ 9
# 15	228 ÷ 6	273 ÷ 7	248 ÷ 8	270 ÷ 9
# 16	186 ÷ 6	210 ÷ 7	304 ÷ 8	351 ÷ 9
# 17	198 ÷ 6	224 ÷ 7	288 ÷ 8	333 ÷ 9
# 18	210 ÷ 6	238 ÷ 7	272 ÷ 8	315 ÷ 9
# 19	222 ÷ 6	252 ÷ 7	256 ÷ 8	297 ÷ 9
# 20	234 ÷ 6	266 ÷ 7	240 ÷ 8	279 ÷ 9

Grade 4 Chapter 3 Lesson 6
Division numbers with quotient from 40 to 49
Divide numbers in column 6
Divide numbers

	a	b	c	d
# 1	$80 \div 2$	$123 \div 3$	$196 \div 4$	$240 \div 5$
# 2	$84 \div 2$	$129 \div 3$	$188 \div 4$	$230 \div 5$
# 3	$88 \div 2$	$135 \div 3$	$180 \div 4$	$220 \div 5$
# 4	$92 \div 2$	$141 \div 3$	$172 \div 4$	$210 \div 5$
# 5	$96 \div 2$	$147 \div 3$	$164 \div 4$	$200 \div 5$
# 6	$82 \div 2$	$120 \div 3$	$192 \div 4$	$245 \div 5$
# 7	$86 \div 2$	$126 \div 3$	$184 \div 4$	$235 \div 5$
# 8	$90 \div 2$	$132 \div 3$	$176 \div 4$	$225 \div 5$
# 9	$94 \div 2$	$138 \div 3$	$168 \div 4$	$215 \div 5$
# 10	$98 \div 2$	$144 \div 3$	$160 \div 4$	$205 \div 5$

Divide numbers in column

$$
\begin{array}{r}
49 \\
9\overline{)441} \\
\underline{-36} \\
81 \\
\underline{-81} \\
0
\end{array}
$$

1) $9 \times 4 = 36$
2) $44 - 36 = 8$
3) Bring down the 1
4) $9 \times 9 = 81$
5) $81 - 81 = 0$

Divide numbers

	a	b	c	d
# 11	$240 \div 6$	$287 \div 7$	$392 \div 8$	$432 \div 9$
# 12	$252 \div 6$	$301 \div 7$	$376 \div 8$	$414 \div 9$
# 13	$264 \div 6$	$315 \div 7$	$360 \div 8$	$396 \div 9$
# 14	$276 \div 6$	$329 \div 7$	$344 \div 8$	$378 \div 9$
# 15	$288 \div 6$	$343 \div 7$	$328 \div 8$	$360 \div 9$
# 16	$246 \div 6$	$280 \div 7$	$384 \div 8$	$441 \div 9$
# 17	$258 \div 6$	$294 \div 7$	$368 \div 8$	$423 \div 9$
# 18	$270 \div 6$	$308 \div 7$	$352 \div 8$	$405 \div 9$
# 19	$282 \div 6$	$322 \div 7$	$336 \div 8$	$387 \div 9$
# 20	$294 \div 6$	$336 \div 7$	$320 \div 8$	$369 \div 9$

Grade 4 Chapter 3 Lesson 7
Division numbers with quotient from 50 to 59
Divide numbers in column 7

```
   59
5 |295
  - 25
    45
  - 45
     0
```

1) 5×5 = 25
2) 29 – 25 = 4
3) Bring down the 5
4) 5×9 = 45
5) 45 – 45 = 0

Divide numbers

	a	b	c	d
# 1	100 ÷ 2	153 ÷ 3	236 ÷ 4	290 ÷ 5
# 2	104 ÷ 2	159 ÷ 3	228 ÷ 4	280 ÷ 5
# 3	108 ÷ 2	165 ÷ 3	220 ÷ 4	270÷ 5
# 4	112 ÷ 2	171 ÷ 3	212 ÷ 4	260 ÷ 5
# 5	116 ÷ 2	177 ÷ 3	204 ÷ 4	250 ÷ 5
# 6	102 ÷ 2	150 ÷ 3	232 ÷ 4	295 ÷ 5
# 7	106 ÷ 2	156 ÷ 3	224 ÷ 4	285 ÷ 5
# 8	110 ÷ 2	162 ÷ 3	216 ÷ 4	275 ÷ 5
# 9	114 ÷ 2	168 ÷ 3	208 ÷ 4	265 ÷ 5
# 10	118 ÷ 2	174 ÷ 3	200 ÷ 4	255 ÷ 5

Divide numbers in column

```
   59
9 |531
  - 45
    81
  - 81
     0
```

1) 9 × 5 = 45
2) 53 – 45 = 8
3) Bring down the 1
4) 9 × 9 = 81
5) 81 – 81 = 0

Divide numbers

	a	b	c	d
# 11	300 ÷ 6	357 ÷ 7	472 ÷ 8	522 ÷ 9
# 12	312 ÷ 6	371 ÷ 7	456 ÷ 8	504 ÷ 9
# 13	324 ÷ 6	385 ÷ 7	440 ÷ 8	486 ÷ 9
# 14	336 ÷ 6	399 ÷ 7	424 ÷ 8	468 ÷ 9
# 15	348 ÷ 6	413 ÷ 7	408 ÷ 8	450 ÷ 9
# 16	306 ÷ 6	350 ÷ 7	464 ÷ 8	531 ÷ 9
# 17	318 ÷ 6	364 ÷ 7	448 ÷ 8	513 ÷ 9
# 18	330 ÷ 6	378 ÷ 7	432 ÷ 8	495 ÷ 9
# 19	342 ÷ 6	392 ÷ 7	416 ÷ 8	477 ÷ 9
# 20	354 ÷ 6	406 ÷ 7	400 ÷ 8	459 ÷ 9

Grade 4 Chapter 3 Lesson 8
Division numbers with quotient from 60 – 69
Divide numbers in column 8

69	1) $5 \times 6 = 30$
5 ⟌345	2) $34 - 30 = 4$
- 30	3) Bring down the 5
45	4) $5 \times 9 = 45$
- 45	5) $45 - 45 = 0$
0	

Divide numbers

	a	b	c	d
# 1	120 ÷ 2	183 ÷ 3	276 ÷ 4	340 ÷ 5
# 2	124 ÷ 2	189 ÷ 3	268 ÷ 4	330 ÷ 5
# 3	128 ÷ 2	195 ÷ 3	260 ÷ 4	320 ÷ 5
# 4	132 ÷ 2	201 ÷ 3	252 ÷ 4	310 ÷ 5
# 5	136 ÷ 2	207 ÷ 3	244 ÷ 4	300 ÷ 5
# 6	122 ÷ 2	180 ÷ 3	272 ÷ 4	345 ÷ 5
# 7	126 ÷ 2	186 ÷ 3	264 ÷ 4	335 ÷ 5
# 8	130 ÷ 2	192 ÷ 3	256 ÷ 4	325 ÷ 5
# 9	134 ÷ 2	198 ÷ 3	248 ÷ 4	315 ÷ 5
# 10	138 ÷ 2	204 ÷ 3	240 ÷ 4	305 ÷ 5

Divide numbers in column

69	1) $9 \times 6 = 54$
9 ⟌621	2) $62 - 54 = 8$
- 54	3) Bring down the 1
81	4) $9 \times 9 = 81$
- 81	5) $81 - 81 = 0$
0	

Divide numbers

	a	b	c	d
# 11	360 ÷ 6	427 ÷ 7	552 ÷ 8	612 ÷ 9
# 12	372 ÷ 6	441 ÷ 7	536 ÷ 8	594 ÷ 9
# 13	384 ÷ 6	455 ÷ 7	520 ÷ 8	576 ÷ 9
# 14	396 ÷ 6	469 ÷ 7	504 ÷ 8	558 ÷ 9
# 15	408 ÷ 6	483 ÷ 7	488 ÷ 8	540 ÷ 9
# 16	366 ÷ 6	420 ÷ 7	544 ÷ 8	621 ÷ 9
# 17	378 ÷ 6	434 ÷ 7	528 ÷ 8	603 ÷ 9
# 18	390 ÷ 6	448 ÷ 7	512 ÷ 8	585 ÷ 9
# 19	402 ÷ 6	462 ÷ 7	496 ÷ 8	567 ÷ 9
# 20	414 ÷ 6	476 ÷ 7	480 ÷ 8	549 ÷ 9

Grade 4 Chapter 3 Lesson 9
Division numbers with quotient from 70 – 79
Divide numbers in column 9
Divide numbers

	a	b	c	d
# 1	140 ÷ 2	213 ÷ 3	316 ÷ 4	390 ÷ 5
# 2	144 ÷ 2	219 ÷ 3	308 ÷ 4	380 ÷ 5
# 3	148 ÷ 2	225 ÷ 3	300 ÷ 4	370 ÷ 5
# 4	152 ÷ 2	231 ÷ 3	292 ÷ 4	360 ÷ 5
# 5	156 ÷ 2	237 ÷ 3	284 ÷ 4	350 ÷ 5
# 6	142 ÷ 2	210 ÷ 3	312 ÷ 4	395 ÷ 5
# 7	146 ÷ 2	216 ÷ 3	304 ÷ 4	385 ÷ 5
# 8	150 ÷ 2	222 ÷ 3	296 ÷ 4	375 ÷ 5
# 9	154 ÷ 2	228 ÷ 3	288 ÷ 4	365 ÷ 5
# 10	158 ÷ 2	234 ÷ 3	280 ÷ 4	355 ÷ 5

Divide numbers in column

```
   79
9|711
 -63
   81
 - 81
    0
```

1) 9× 7 = 63
2) 71 – 63 = 8
3) Bring down the 1
4) 9 × 9 = 81
5) 81 – 81 = 0

Divide numbers

	a	b	c	d
# 11	420 ÷ 6	497 ÷ 7	632 ÷ 8	702 ÷ 9
# 12	432 ÷ 6	511 ÷ 7	616 ÷ 8	684 ÷ 9
# 13	444 ÷ 6	525 ÷ 7	600 ÷ 8	666 ÷ 9
# 14	456 ÷ 6	539 ÷ 7	584 ÷ 8	648 ÷ 9
# 15	468 ÷ 6	553 ÷ 7	568 ÷ 8	630 ÷ 9
# 16	426 ÷ 6	490 ÷ 7	624 ÷ 8	711 ÷ 9
# 17	438 ÷ 6	504 ÷ 7	608 ÷ 8	693 ÷ 9
# 18	450 ÷ 6	518 ÷ 7	592 ÷ 8	675 ÷ 9
# 19	462 ÷ 6	532 ÷ 7	576 ÷ 8	657 ÷ 9
# 20	474 ÷ 6	546 ÷ 7	560 ÷ 8	639 ÷ 9

Grade 4 Chapter 3 Lesson 10

Division numbers with quotient from 80 to 89

Divide numbers in column 10

Divide numbers

	a	b	c	d
# 1	160 ÷ 2	243 ÷ 3	356 ÷ 4	440 ÷ 5
# 2	164 ÷ 2	249 ÷ 3	348 ÷ 4	430 ÷ 5
# 3	168 ÷ 2	255 ÷ 3	340 ÷ 4	420 ÷ 5
# 4	172 ÷ 2	261 ÷ 3	332 ÷ 4	410 ÷ 5
# 5	176 ÷ 2	267 ÷ 3	324 ÷ 4	400 ÷ 5
# 6	162 ÷ 2	240 ÷ 3	352 ÷ 4	445 ÷ 5
# 7	166 ÷ 2	246 ÷ 3	344 ÷ 4	435 ÷ 5
# 8	170 ÷ 2	252 ÷ 3	336 ÷ 4	425 ÷ 5
# 9	174 ÷ 2	258 ÷ 3	328 ÷ 4	415 ÷ 5
# 10	178 ÷ 2	264 ÷ 3	320 ÷ 4	405 ÷ 5

Divide numbers in column

$$\begin{array}{r} 89 \\ 9\overline{)801} \\ \underline{-72} \\ 81 \\ \underline{-81} \\ 0 \end{array}$$

1) $9 \times 8 = 72$
2) $80 - 72 = 8$
3) Bring down the 1
4) $9 \times 9 = 81$
5) $81 - 81 = 0$

Divide numbers

	a	b	c	d
# 11	480 ÷ 6	567 ÷ 7	712 ÷ 8	792 ÷ 9
# 12	492 ÷ 6	581 ÷ 7	696 ÷ 8	774 ÷ 9
# 13	504 ÷ 6	595 ÷ 7	680 ÷ 8	756 ÷ 9
# 14	516 ÷ 6	609 ÷ 7	664 ÷ 8	738 ÷ 9
# 15	528 ÷ 6	623 ÷ 7	648 ÷ 8	720 ÷ 9
# 16	486 ÷ 6	560 ÷ 7	704 ÷ 8	801 ÷ 9
# 17	498 ÷ 6	574 ÷ 7	688 ÷ 8	783 ÷ 9
# 18	510 ÷ 6	588 ÷ 7	672 ÷ 8	765 ÷ 9
# 19	522 ÷ 6	602 ÷ 7	656 ÷ 8	747 ÷ 9
# 20	534 ÷ 6	616 ÷ 7	640 ÷ 8	729 ÷ 9

Grade 4 Chapter 3 Lesson 11

Division numbers with quotient from 90 to 99

Divide numbers in column 11

Divide numbers

	a	b	c	d
# 1	180 ÷ 2	273 ÷ 3	396 ÷ 4	490 ÷ 5
# 2	184 ÷ 2	279 ÷ 3	388 ÷ 4	480 ÷ 5
# 3	188 ÷ 2	285 ÷ 3	380 ÷ 4	470 ÷ 5
# 4	192 ÷ 2	291 ÷ 3	372 ÷ 4	460 ÷ 5
# 5	196 ÷ 2	297 ÷ 3	364 ÷ 4	450 ÷ 5
# 6	182 ÷ 2	270 ÷ 3	392 ÷ 4	495 ÷ 5
# 7	186 ÷ 2	276 ÷ 3	384 ÷ 4	485 ÷ 5
# 8	190 ÷ 2	282 ÷ 3	376 ÷ 4	475 ÷ 5
# 9	194 ÷ 2	288 ÷ 3	368 ÷ 4	465 ÷ 5
# 10	198 ÷ 2	294 ÷ 3	360 ÷ 4	455 ÷ 5

Divide numbers in column

```
   97
8|776
 -72
   56
  -56
    0
```

1) $8 \times 9 = 72$
2) $77 - 72 = 5$
3) Bring down the 6
4) $8 \times 7 = 56$
5) $56 - 56 = 0$

Divide numbers

	a	b	c	d
# 11	540 ÷ 6	637 ÷ 7	792 ÷ 8	882 ÷ 9
# 12	552 ÷ 6	651 ÷ 7	776 ÷ 8	864 ÷ 9
# 13	564 ÷ 6	665 ÷ 7	760 ÷ 8	846 ÷ 9
# 14	576 ÷ 6	679 ÷ 7	744 ÷ 8	828 ÷ 9
# 15	588 ÷ 6	693 ÷ 7	728 ÷ 8	810 ÷ 9
# 16	546 ÷ 6	630 ÷ 7	784 ÷ 8	891 ÷ 9
# 17	558 ÷ 6	644 ÷ 7	768 ÷ 8	873 ÷ 9
# 18	570 ÷ 6	658 ÷ 7	752 ÷ 8	855 ÷ 9
# 19	582 ÷ 6	672 ÷ 7	736 ÷ 8	837 ÷ 9
# 20	594 ÷ 6	686 ÷ 7	720 ÷ 8	819 ÷ 9

Grade 4 Chapter 3 Lesson 12

Division numbers

Review 1 A

Divide the numbers in column

# 1	490 ÷5	# 11	531 ÷ 9
# 2	522 ÷ 6	# 12	544 ÷ 8
# 3	532 ÷ 7	# 13	525 ÷ 7
# 4	520 ÷ 8	# 14	516 ÷ 6
# 5	486 ÷ 9	# 15	475 ÷ 5
# 6	485 ÷ 5	# 16	504 ÷ 9
# 7	516 ÷ 6	# 17	536 ÷ 8
# 8	525 ÷ 7	# 18	546 ÷ 7
# 9	512 ÷ 8	# 19	534 ÷ 6
# 10	477 ÷ 9	# 20	480 ÷ 5

Grade 4 Chapter 3 Lesson 12

Review 1 B

Divide the numbers in column

# 1	395 ÷5	# 11	472 ÷ 8
# 2	602 ÷ 7	# 12	552 ÷ 8
# 3	608 ÷ 8	# 13	450 ÷ 6
# 4	536 ÷ 8	# 14	696 ÷ 8
# 5	504 ÷ 9	# 15	582 ÷ 6
# 6	390 ÷ 5	# 16	448 ÷ 8
# 7	595 ÷ 7	# 17	603 ÷ 9
# 8	450 ÷ 6	# 18	702 ÷ 9
# 9	576 ÷ 9	# 19	623 ÷ 7
# 10	424 ÷ 8	# 20	445 ÷ 5

Grade 4 Chapter 3 Lesson 13

Review 2 A

Divide the numbers in column

# 1	480 ÷ 5	# 11	513 ÷ 9
# 2	510 ÷ 6	# 12	528 ÷ 8
# 3	518 ÷ 7	# 13	504 ÷ 7
# 4	504 ÷ 8	# 14	504 ÷ 6
# 5	468 ÷ 9	# 15	465 ÷ 5
# 6	475 ÷ 5	# 16	486 ÷ 9
# 7	504 ÷ 6	# 17	520 ÷ 8
# 8	511 ÷ 7	# 18	532 ÷ 7
# 9	496 ÷ 8	# 19	522 ÷ 6
# 10	459 ÷ 9	# 20	470 ÷ 5

Grade 4 Chapter 3 Lesson 13

Review 2 B

Divide the numbers in column

# 1	$385 \div 5$	# 11	$456 \div 8$
# 2	$588 \div 7$	# 12	$536 \div 8$
# 3	$592 \div 8$	# 13	$438 \div 6$
# 4	$520 \div 8$	# 14	$680 \div 8$
# 5	$486 \div 9$	# 15	$570 \div 6$
# 6	$380 \div 5$	# 16	$432 \div 8$
# 7	$581 \div 7$	# 17	$585 \div 9$
# 8	$438 \div 6$	# 18	$684 \div 9$
# 9	$558 \div 9$	# 19	$609 \div 7$
# 10	$408 \div 8$	# 20	$430 \div 5$

Grade 4 Chapter 3 Lesson 14
Summary 3

1) Multiplication numbers: for memory (x, x, any)
To multiply numbers, work in any order
$9 \times 8 \times 7 = 72 \times 7 = 504$ $9 \times 8 \times 7 = 9 \times 56 = 504$

2) Work from left to right: for memory (L → R)

a) Multiplication and Division numbers: (× or ÷)
To multiply and divide numbers, work from left to right (L → R)
$9 \times 6 \div 3 = 54 \div 3 = 18$ $4 \times 81 \div 9 = 324 \div 9 = 36$

b) Division and Multiplication numbers: (÷ or ×)
To divide and multiply numbers, work from left to right (L → R)
$24 \div 6 \times 2 = 4 \times 2 = 8$ $81 \div 9 \times 4 = 9 \times 4 = 36$

c) Division numbers: (÷, ÷)
To divide numbers, work from left to right (L → R)
$24 \div 4 \div 2 = 6 \div 2 = 3$ $810 \div 9 \div 3 = 90 \div 9 = 10$

3) Do Multiplication or Division before Addition or Subtraction
for memory: (× or ÷) before (+ or -)
a) Multiplication or Addition numbers: 1) × 2) +
To multiply or add numbers, always first multiply than add
$9 \times 6 + 9 = 54 + 9 = 63$ $9 + 9 \times 9 = 9 + 81 = 90$

b) Multiplication or Subtraction numbers: 1) × 2) −
To multiply or subtract numbers, always first multiply than subtract
$9 \times 9 - 9 = 81 - 9 = 72$ $90 - 9 \times 9 = 90 - 81 = 9$

c) Division or Addition numbers: 1) ÷ 2) +
To divide or add numbers, always first divide than add
$50 \div 5 + 39 = 10 + 39 = 49$ $88 + 171 \div 9 = 88 + 19 = 107$

d) Division or Subtraction numbers: 1) ÷ 2) −
To divide or subtract numbers, always first divide than subtract
$108 \div 6 - 9 = 18 - 9 = 9$ $500 - 126 \div 7 = 500 - 18 = 482$

Grade 4 Chapter 3 Lesson 14

Multiplication numbers

To multiply numbers, work in any order

$5\times6\times10=30\times10=300$

$5\times6\times10 = 5\times60 = 300$

Multiply numbers

	a	b	c
# 1	2 × 2 × 1	3 × 4 × 1	4 × 6 × 1
# 2	2 × 2 × 2	3 × 4 × 2	4 × 6 × 2
# 3	2 × 2 × 3	3 × 4 × 3	4 × 6 × 3
# 4	2 × 2 × 4	3 × 4 × 4	4 × 6 × 4
# 5	2 × 2 × 5	3 × 4 × 5	4 × 6 × 5
# 6	2 × 2 × 6	3 × 4 × 6	4 × 6 × 6
# 7	2 × 2 × 7	3 × 4 × 7	4 × 6 × 7
# 8	2 × 2 × 8	3 × 4 × 8	4 × 6 × 8
# 9	2 × 2 × 9	3 × 4 × 9	4 × 6 × 9
# 10	2 × 2 × 10	3 × 4 × 10	4 × 6 × 10

To multiply numbers, work in any order

$5\times8\times10 =40\times10=400$

$5\times8\times10 = 5\times80 = 400$

Multiply numbers

	a	b	c
# 11	5 × 3 × 1	6 × 5 × 1	7 × 7 × 1
# 12	5 × 3 × 2	6 × 5 × 2	7 × 7 × 2
# 13	5 × 3 × 3	6 × 5 × 3	7 × 7× 3
# 14	5 × 3 × 4	6 × 5 × 4	7 × 7 × 4
# 15	5 × 3 × 5	6 × 5 × 5	7 × 7 × 5
# 16	5 × 3 × 6	6 × 5 × 6	7 × 7 × 6
# 17	5 × 3 × 7	6 × 5 × 7	7 × 7 × 7
# 18	5 × 3 × 8	6 × 5 × 8	7 × 7 × 8
# 19	5 × 3 × 9	6 × 5 × 9	7 × 7 × 9
# 20	5 × 3 × 10	6 × 5 × 10	7 × 7 × 10

Grade 4 Chapter 3 Lesson 14

Multiplication and Division numbers

To multiply and divide numbers work from left to right

$5 \times 60 \div 6 = 300 \div 6 = 50$

Multiply numbers

	a	b	c
# 21	$2 \times 2 \div 2$	$3 \times 4 \div 4$	$4 \times 6 \div 6$
# 22	$2 \times 4 \div 2$	$3 \times 8 \div 4$	$4 \times 12 \div 6$
# 23	$2 \times 6 \div 2$	$3 \times 12 \div 4$	$4 \times 18 \div 6$
# 24	$2 \times 8 \div 2$	$3 \times 16 \div 4$	$4 \times 24 \div 6$
# 25	$2 \times 10 \div 2$	$3 \times 20 \div 4$	$4 \times 30 \div 6$
# 26	$2 \times 12 \div 2$	$3 \times 24 \div 4$	$4 \times 36 \div 6$
# 27	$2 \times 14 \div 2$	$3 \times 28 \div 4$	$4 \times 42 \div 6$
# 28	$2 \times 16 \div 2$	$3 \times 32 \div 4$	$4 \times 48 \div 6$
# 29	$2 \times 18 \div 2$	$3 \times 36 \div 4$	$4 \times 54 \div 6$
# 30	$2 \times 20 \div 2$	$3 \times 40 \div 4$	$4 \times 60 \div 6$

Multiplication and Division numbers

To multiply and divide numbers work from left to right

$5 \times 80 \div 8 = 400 \div 8 = 50$

Multiply numbers

	a	b	c
# 31	$5 \times 3 \div 3$	$6 \times 5 \div 5$	$7 \times 7 \div 7$
# 32	$5 \times 6 \div 3$	$6 \times 10 \div 5$	$7 \times 14 \div 7$
# 33	$5 \times 9 \div 3$	$6 \times 15 \div 5$	$7 \times 21 \div 7$
# 34	$5 \times 12 \div 3$	$6 \times 20 \div 5$	$7 \times 28 \div 7$
# 35	$5 \times 15 \div 3$	$6 \times 25 \div 5$	$7 \times 35 \div 7$
# 36	$5 \times 18 \div 3$	$6 \times 30 \div 5$	$7 \times 42 \div 7$
# 37	$5 \times 21 \div 3$	$6 \times 35 \div 5$	$7 \times 49 \div 7$
# 38	$5 \times 24 \div 3$	$6 \times 40 \div 5$	$7 \times 56 \div 7$
# 39	$5 \times 27 \div 3$	$6 \times 45 \div 5$	$7 \times 63 \div 7$
# 40	$5 \times 30 \div 3$	$6 \times 50 \div 5$	$7 \times 70 \div 7$

Grade 4 Chapter 3 Lesson 15

Division and Multiplication numbers

To divide and multiply numbers work from left to right

$70 \div 7 \times 4 = 10 \times 4 = 40$

Evaluate

	a	b	c
# 1	$2 \div 2 \times 2$	$4 \div 4 \times 3$	$6 \div 6 \times 4$
# 2	$4 \div 2 \times 2$	$8 \div 4 \times 3$	$12 \div 6 \times 4$
# 3	$6 \div 2 \times 2$	$12 \div 4 \times 3$	$18 \div 6 \times 4$
# 4	$8 \div 2 \times 2$	$16 \div 4 \times 3$	$24 \div 6 \times 4$
# 5	$10 \div 2 \times 2$	$20 \div 4 \times 3$	$30 \div 6 \times 4$
# 6	$12 \div 2 \times 2$	$24 \div 4 \times 3$	$36 \div 6 \times 4$
# 7	$14 \div 2 \times 2$	$28 \div 4 \times 3$	$42 \div 6 \times 4$
# 8	$16 \div 2 \times 2$	$32 \div 4 \times 3$	$48 \div 6 \times 4$
# 9	$18 \div 2 \times 2$	$36 \div 4 \times 3$	$54 \div 6 \times 4$
# 10	$20 \div 2 \times 2$	$40 \div 4 \times 3$	$60 \div 6 \times 4$

Division and Multiplication numbers

To divide and multiply numbers work from left to right

$80 \div 8 \times 8 = 10 \times 8 = 80$

Evaluate

	a	b	c
# 11	$3 \div 3 \times 5$	$5 \div 5 \times 6$	$7 \div 7 \times 7$
# 12	$6 \div 3 \times 5$	$10 \div 5 \times 6$	$14 \div 7 \times 7$
# 13	$9 \div 3 \times 5$	$15 \div 5 \times 6$	$21 \div 7 \times 7$
# 14	$12 \div 3 \times 5$	$20 \div 5 \times 6$	$28 \div 7 \times 7$
# 15	$15 \div 3 \times 5$	$25 \div 5 \times 6$	$35 \div 7 \times 7$
# 16	$18 \div 3 \times 5$	$30 \div 5 \times 6$	$42 \div 7 \times 7$
# 17	$21 \div 3 \times 5$	$35 \div 5 \times 6$	$49 \div 7 \times 7$
# 18	$24 \div 3 \times 5$	$40 \div 5 \times 6$	$56 \div 7 \times 7$
# 19	$27 \div 3 \times 5$	$45 \div 5 \times 6$	$63 \div 7 \times 7$
# 20	$30 \div 3 \times 5$	$50 \div 5 \times 6$	$70 \div 7 \times 7$

Grade 4 Chapter 3 Lesson 15

Division numbers

To divide numbers, work from left to right

$640 \div 8 \div 8 = 80 \div 8 = 10$

Divide numbers

	a	b	c
# 21	4 ÷ 2 ÷ 2	16 ÷ 4 ÷ 4	36 ÷ 6 ÷ 6
# 22	8 ÷ 2 ÷ 2	32 ÷ 4 ÷ 4	72 ÷ 6÷ 6
# 23	12 ÷ 2 ÷ 2	48 ÷ 4 ÷ 4	108 ÷ 6 ÷ 6
# 24	16 ÷ 2 ÷ 2	64 ÷ 4 ÷ 4	144 ÷ 6 ÷ 6
# 25	20 ÷ 2 ÷ 2	80 ÷ 4 ÷ 4	180 ÷ 6 ÷ 6
# 26	24 ÷ 2 ÷ 2	96 ÷ 4 ÷ 4	216 ÷ 6 ÷ 6
# 27	28 ÷ 2 ÷ 2	112 ÷ 4 ÷ 4	252 ÷ 6 ÷ 6
# 28	32 ÷ 2 ÷ 2	128 ÷ 4÷ 4	288 ÷ 6 ÷ 6
# 29	36 ÷ 2 ÷ 2	144 ÷ 4 ÷ 4	324 ÷ 6 ÷ 6
# 30	40 ÷ 2 ÷ 2	160 ÷ 4 ÷ 4	360 ÷ 6 ÷ 6

Divide numbers

To divide numbers, work from left to right

$810 \div 9 \div 9 = 90 \div 9 = 10$

Divide numbers

	a	b	c
# 31	9 ÷ 3 ÷ 3	25÷5 ÷ 5	49 ÷ 7÷7
# 32	18 ÷ 3 ÷3	50 ÷5÷5	98 ÷7÷7
# 33	27÷3 ÷3	75 ÷5÷5	147 ÷7÷7
# 34	36 ÷3÷3	100 ÷5÷5	196 ÷7÷7
# 35	45 ÷3÷3	125 ÷5÷5	245÷7 ÷7
# 36	54 ÷3÷3	150 ÷5÷5	294 ÷7 ÷7
# 37	63 ÷3÷3	175÷5÷5	343÷7÷ 7
# 38	72 ÷3÷3	200÷5÷5	392 ÷7÷ 7
# 39	81 ÷3÷3	225÷5 ÷5	441 ÷7÷ 7
# 40	90 ÷3÷3	250÷5÷5	490 ÷7÷ 7

Grade 4 Chapter 3 Lesson 16

Division and Addition numbers

To divide and add numbers always first divide than add

$$50 \div 5 + 10 = 10 + 10 = 20$$

Evaluate

	a	b	c	d
# 1	2÷2 + 0	3÷3 + 0	4÷4 + 0	5÷5 + 0
# 2	4÷2 + 1	6÷3 + 1	8÷4 + 1	10÷5 +1
# 3	6÷2 + 2	9÷3 + 2	12÷4 + 2	15÷5 + 2
# 4	8÷2 + 3	12÷3 +3	16÷4 + 3	20÷5 + 3
# 5	10÷2 + 4	15÷3 +4	20÷4 + 4	25÷5 + 4
# 6	12÷2 + 5	18÷3 +5	24÷4 + 5	30÷5 + 5
# 7	14÷2 + 6	21÷3 +6	28÷4 + 6	35÷5 + 6
# 8	16÷2 + 7	24÷3 +7	32÷4 + 7	40÷5 + 7
# 9	18÷2 + 8	27÷3 +8	36÷4 + 8	45÷5 + 8
# 10	20÷2 + 9	30÷3 +9	40÷4 + 9	50÷5 + 9

Divide and Add

$$10 \overset{2}{+} 90 \overset{1}{\div} 9 = 10 + 10 = 20$$

Evaluate

	a	b	c	d
# 11	0 + 6÷6	0 + 7÷7	0 + 8÷8	0 + 9÷9
# 12	1 + 12÷6	1 + 14÷7	1 + 16÷8	1 + 18÷9
# 13	2 + 18÷6	2 + 21÷7	2 + 24÷8	2 + 27÷9
# 14	3 + 24÷6	3 + 28÷7	3 + 32÷8	3 + 36÷9
# 15	4 + 30÷6	4 + 35÷7	4 + 40÷8	4 + 45÷9
# 16	5 + 36÷6	5 + 42÷7	5 + 48÷8	5 + 54÷9
# 17	6 + 42÷6	6 + 49÷7	6 + 56÷8	6 + 63÷9
# 18	7 + 48÷6	7 + 56÷7	7 + 64÷8	7 + 72÷9
# 19	8 + 54÷6	8 + 63÷7	8 + 72÷8	8 +81÷9
# 20	9 + 60÷6	9 + 70÷7	9 + 80÷8	9 + 90÷9

Grade 4 Chapter 3 Lesson 16
Division and Subtraction numbers
To divide and subtract numbers always first divide than subtract

$$80 \div 5 - 10 = 16 - 10 = 6$$

Evaluate

	a	b	c	d
# 21	$22 \div 2 - 1$	$24 \div 3 - 2$	$24 \div 4 - 3$	$25 \div 5 - 4$
# 22	$24 \div 2 - 1$	$27 \div 3 - 2$	$28 \div 4 - 3$	$30 \div 5 - 4$
# 23	$26 \div 2 - 1$	$30 \div 3 - 2$	$32 \div 4 - 3$	$35 \div 5 - 4$
# 24	$28 \div 2 - 1$	$33 \div 3 - 2$	$36 \div 4 - 3$	$40 \div 5 - 4$
# 25	$30 \div 2 - 1$	$36 \div 3 - 2$	$40 \div 4 - 3$	$45 \div 5 - 4$
# 26	$32 \div 2 - 1$	$39 \div 3 - 2$	$44 \div 4 - 3$	$50 \div 5 - 4$
# 27	$34 \div 2 - 1$	$42 \div 3 - 2$	$48 \div 4 - 3$	$55 \div 5 - 4$
# 28	$36 \div 2 - 1$	$45 \div 3 - 2$	$52 \div 4 - 3$	$60 \div 5 - 4$
# 29	$38 \div 2 - 1$	$48 \div 3 - 2$	$56 \div 4 - 3$	$65 \div 5 - 4$
# 30	$40 \div 2 - 1$	$51 \div 3 - 2$	$60 \div 4 - 3$	$70 \div 5 - 4$

Divide and Subtract numbers

2 1

$$50 - 20 \div 4 = 50 - 5 = 45$$

Evaluate

	a	b	c	d
# 31	$18 - 6 \div 6$	$21 - 7 \div 7$	$24 - 8 \div 8$	$27 - 9 \div 9$
# 32	$24 - 12 \div 6$	$28 - 14 \div 7$	$32 - 16 \div 8$	$36 - 18 \div 9$
# 33	$30 - 18 \div 6$	$35 - 21 \div 7$	$36 - 24 \div 8$	$45 - 27 \div 9$
# 34	$36 - 24 \div 6$	$42 - 28 \div 7$	$48 - 32 \div 8$	$54 - 36 \div 9$
# 35	$42 - 30 \div 6$	$49 - 35 \div 7$	$56 - 40 \div 8$	$63 - 45 \div 9$
# 36	$48 - 36 \div 6$	$56 - 42 \div 7$	$64 - 48 \div 8$	$72 - 54 \div 9$
# 37	$54 - 42 \div 6$	$63 - 49 \div 7$	$72 - 56 \div 8$	$81 - 63 \div 9$
# 38	$60 - 48 \div 6$	$70 - 56 \div 7$	$80 - 64 \div 8$	$90 - 72 \div 9$
# 39	$66 - 54 \div 6$	$77 - 63 \div 7$	$88 - 72 \div 8$	$99 - 81 \div 9$
# 40	$72 - 60 \div 6$	$84 - 70 \div 7$	$96 - 80 \div 8$	$108 - 90 \div 9$

Grade 4 Chapter 3 Lesson 17

Review 4 A

Evaluate

# 1	$9 \times 2 \times 7$	# 11	$9 \times 3 \times 6$
# 2	$9 \times 18 \div 6$	# 12	$8 \times 27 \div 3$
# 3	$54 \div 9 \times 4$	# 13	$63 \div 7 \times 5$
# 4	$819 \div 7 \div 9$	# 14	$828 \div 4 \div 9$
# 5	$100 + 7 \times 7$	# 15	$110 + 6 \times 6$
# 6	$80 \times 9 - 8$	# 16	$90 \times 8 - 9$
# 7	$140 - 8 \times 5$	# 17	$150 - 8 \times 6$
# 8	$800 \div 8 - 6$	# 18	$840 \div 8 - 6$
# 9	$666 + 333 \div 3$	# 19	$440 + 200 \div 5$
# 10	$840 - 240 \div 8$	# 20	$870 - 350 \div 5$

Grade 4 Chapter 3 Lesson 17

Review 4 B

Evaluate

# 1	$9 \times 4 \times 5$	# 11	$9 \times 5 \times 4$
# 2	$6 \times 36 \div 9$	# 12	$7 \times 45 \div 9$
# 3	$63 \div 9 \times 3$	# 13	$54 \div 9 \times 2$
# 4	$108 \div 9 \div 3$	# 14	$972 \div 9 \div 2$
# 5	$104 + 8 \times 7$	# 15	$101 + 7 \times 7$
# 6	$90 \times 8 - 20$	# 16	$80 \times 7 - 60$
# 7	$120 - 12 \times 5$	# 17	$170 - 14 \times 5$
# 8	$960 \div 8 - 20$	# 18	$900 \div 9 - 60$
# 9	$260 + 280 \div 7$	# 19	$270 + 180 \div 6$
# 10	$350 - 400 \div 8$	# 20	$450 - 450 \div 9$

Grade 4 Chapter 3 Lesson 18

Four operations with numbers

Test 4 A

Evaluate

# 1	$9 \times 2 \times 9$	# 11	$9 \times 3 \times 8$
# 2	$9 \times 18 \div 9$	# 12	$8 \times 27 \div 9$
# 3	$54 \div 9 \times 5$	# 13	$63 \div 9 \times 4$
# 4	$819 \div 9 \div 7$	# 14	$828 \div 9 \div 4$
# 5	$100 + 9 \times 9$	# 15	$110 + 8 \times 8$
# 6	$80 \times 9 - 6$	# 16	$90 \times 8 - 7$
# 7	$140 - 9 \times 5$	# 17	$150 - 9 \times 6$
# 8	$840 \div 8 - 6$	# 18	$800 \div 8 - 6$
# 9	$666 + 333 \div 9$	# 19	$440 + 200 \div 8$
# 10	$840 - 240 \div 6$	# 20	$850 - 350 \div 7$

Grade 4 Chapter 3 Lesson 18

Four operations with numbers

Test 4 B

Evaluate

# 1	$9 \times 4 \times 7$	# 11	$9 \times 5 \times 6$
# 2	$7 \times 36 \div 9$	# 12	$6 \times 45 \div 9$
# 3	$72 \div 9 \times 3$	# 13	$81 \div 9 \times 2$
# 4	$837 \div 9 \div 3$	# 14	$846 \div 9 \div 2$
# 5	$120 + 8 \times 7$	# 15	$130 + 7 \times 7$
# 6	$90 \times 8 - 5$	# 16	$80 \times 9 - 6$
# 7	$160 - 9 \times 7$	# 17	$170 - 9 \times 8$
# 8	$960 \div 8 - 6$	# 18	$900 \div 9 - 6$
# 9	$210 + 280 \div 7$	# 19	$300 + 180 \div 6$
# 10	$800 - 400 \div 8$	# 20	$900 - 450 \div 9$

Grade 4 Answers Chapter 3

Lesson 1

	a	b	c	d		a	b	c	d
# 1	1	6	1	6	# 6	1	6	1	6
# 2	2	7	2	7	# 7	2	7	2	7
# 3	3	8	3	8	# 8	3	8	3	8
# 4	4	9	4	9	# 9	4	9	4	9
# 5	5	10	5	10	# 10	5	10	5	10

Lesson 2

	a	b	c	d		a	b	c	d
# 1	0	1	9	0	# 6	9	8	1	2
# 2	2	3	8	7	# 7	6	7	3	4
# 3	5	4	6	5	# 8	4	5	5	6
# 4	7	6	4	3	# 9	3	2	7	8
# 5	8	9	2	1	# 10	1	0	0	9

Lesson 2

# 1	×1	2	3	4	5	6	7	8	9
# 2	2	4	6	8	10	12	14	16	18
# 3	3	6	9	12	15	18	21	24	27
# 4	4	8	12	16	20	24	28	32	36
# 5	5	10	15	20	25	30	35	40	45
# 6	6	12	18	24	30	36	42	48	54
# 7	7	14	21	28	35	42	49	56	63
# 8	8	16	24	32	40	48	56	64	72
# 9	9	18	27	36	45	54	63	72	81
#10	10	20	30	40	50	60	70	80	90

Lesson 3	a b c d		a b c d
# 1	10, 11, 19, 18	# 11	10, 11, 19, 18
# 2	12, 13, 17, 16	# 12	12, 13, 17, 16
# 3	14, 15, 15, 14	# 13	14, 15, 15, 14
# 4	16, 17, 13, 12	# 14	16, 17, 13, 12
# 5	18, 19, 11, 10	# 15	18, 19, 11, 10
# 6	11, 10, 18, 19	# 16	11, 10, 18, 19
# 7	13, 12, 16, 17	# 17	13, 12, 16, 17
# 8	15, 14, 14, 15	# 18	15, 14, 14, 15
# 9	17, 16, 12, 13	# 19	17, 16, 12, 13
# 10	19, 18, 10, 11	# 20	19, 18, 10, 11

Grade 4 Answers Chapter 3

Lesson 4

	a b c d		a b c d
# 1	20, 21, 29, 28	# 11	20, 21, 29, 28
# 2	22, 23, 27, 26	# 12	22, 23, 27, 26
# 3	24, 25, 25, 24	# 13	24, 25, 25, 24
# 4	26, 27, 23, 22	# 14	26, 27, 23, 22
# 5	28, 29, 21, 20	# 15	28, 29, 21, 20
# 6	21, 20, 28, 29	# 16	21, 20, 28, 29
# 7	23, 22, 26, 27	# 17	23, 22, 26, 27
# 8	25, 24, 24, 25	# 18	25, 24, 24, 25
# 9	27, 26, 22, 23	# 19	27, 26, 22, 23
# 10	29, 28, 20, 21	# 20	29, 28, 20, 21

Lesson 5

	a b c d		a b c d
# 1	30, 31, 39, 38	# 11	30, 31, 39, 38
# 2	32, 33, 37, 36	# 12	32, 33, 37, 36
# 3	34, 35, 35, 34	# 13	34, 35, 35, 34
# 4	36, 37, 33, 32	# 14	36, 37, 33, 32
# 5	38, 39, 31, 30	# 15	38, 39, 31, 30
# 6	31, 30, 38, 39	# 16	31, 30, 38, 39
# 7	33, 32, 36, 37	# 17	33, 32, 36, 37
# 8	35, 34, 34, 35	# 18	35, 34, 34, 35
# 9	37, 36, 32, 33	# 19	37, 36, 32, 33
# 10	39, 38, 30, 31	# 20	39, 38, 30, 31

Lesson 6

	a b c d		a b c d
# 1	40, 41, 49, 48	# 11	40, 41, 49, 48
# 2	42, 43, 47, 46	# 12	42, 43, 47, 46
# 3	44, 45, 45, 44	# 13	44, 45, 45, 44
# 4	46, 47, 43, 42	# 14	46, 47, 43, 42
# 5	48, 49, 41, 40	# 15	48, 49, 41, 40
# 6	41, 40, 48, 49	# 16	41, 40, 48, 49
# 7	43, 42, 46, 47	# 17	43, 42, 46, 47
# 8	45, 44, 44, 45	# 18	45, 44, 44, 45
# 9	47, 46, 42, 43	# 19	47, 46, 42, 43
# 10	49, 48, 40, 41	# 20	49, 48, 40, 41

Grade 4 Answers Chapter 3

Lesson 7

	a b c d		a b c d
# 1	50, 51, 59, 58	# 11	50, 51, 59, 58
# 2	52, 53, 57, 56	# 12	52, 53, 57, 56
# 3	54, 55, 55, 54	# 13	54, 55, 55, 54
# 4	56, 57, 53, 52	# 14	56, 57, 53, 52
# 5	58, 59, 51, 50	# 15	58, 59, 51, 50
# 6	51, 50, 58, 59	# 16	51, 50, 58, 59
# 7	53, 52, 56, 57	# 17	53, 52, 56, 57
# 8	55, 54, 54, 55	# 18	55, 54, 54, 55
# 9	57, 56, 52, 53	# 19	57, 56, 52, 53
# 10	59, 58, 50, 51	# 20	59, 58, 50, 51

Lesson 8

	a b c d		a b c d
# 1	60, 61, 69, 68	# 11	60, 61, 69, 68
# 2	62, 63, 67, 66	# 12	62, 63, 67, 66
# 3	64, 65, 65, 64	# 13	64, 65, 65, 64
# 4	66, 67, 63, 62	# 14	66, 67, 63, 62
# 5	68, 69, 61, 60	# 15	68, 69, 61, 60
# 6	61, 60, 68, 69	# 16	61, 60, 68, 69
# 7	63, 62, 66, 67	# 17	63, 62, 66, 67
# 8	65, 64, 64, 65	# 18	65, 64, 64, 65
# 9	67, 66, 62, 63	# 19	67, 66, 62, 63
# 10	69, 68, 60, 61	# 20	69, 68, 60, 61

Lesson 9

	a b c d		a b c d
# 1	70, 71, 79, 78	# 11	70, 71, 79, 78
# 2	72, 73, 77, 76	# 12	72, 73, 77, 76
# 3	74, 75, 75, 74	# 13	74, 75, 75, 74
# 4	76, 77, 73, 72	# 14	76, 77, 73, 72
# 5	78, 79, 71, 70	# 15	78, 79, 71, 70
# 6	71, 70, 78, 79	# 16	71, 70, 78, 79
# 7	73, 72, 76, 77	# 17	73, 72, 76, 77
# 8	75, 74, 74, 75	# 18	75, 74, 74, 75
# 9	77, 76, 72, 73	# 19	77, 76, 72, 73
# 10	79, 78, 70, 71	# 20	79, 78, 70, 71

Grade 4 Answers Chapter 3

Lesson 10

	a b c d		a b c d
# 1	80, 81, 89, 88	# 11	80, 81, 89, 88
# 2	82, 83, 87, 86	# 12	82, 83, 87, 86
# 3	84, 85, 85, 84	# 13	84, 85, 85, 84
# 4	86, 87, 83, 82	# 14	86, 87, 83, 82
# 5	88, 89, 81, 80	# 15	88, 89, 81, 80
# 6	81, 80, 88, 89	# 16	81, 80, 88, 89
# 7	83, 82, 86, 87	# 17	83, 82, 86, 87
# 8	85, 84, 84, 85	# 18	85, 84, 84, 85
# 9	87, 86, 82, 83	# 19	87, 86, 82, 83
# 10	89, 88, 80, 81	# 20	89, 88, 80, 81

Lesson 11

	a b c d		a b c d
# 1	90, 91, 99, 98	# 11	90, 91, 99, 98
# 2	92, 93, 97, 96	# 12	92, 93, 97, 96
# 3	94, 95, 95, 94	# 13	94, 95, 95, 94
# 4	96, 97, 93, 92	# 14	96, 97, 93, 92
# 5	98, 99, 91, 90	# 15	98, 99, 91, 90
# 6	91, 90, 98, 99	# 16	91, 90, 98, 99
# 7	93, 92, 96, 97	# 17	93, 92, 96, 97
# 8	95, 94, 94, 95	# 18	95, 94, 94, 95
# 9	97, 96, 92, 93	# 19	97, 96, 92, 93
# 10	99, 98, 90, 91	# 20	99, 98, 90, 91

Lesson 12

Review 1 A

# 1	98	# 6	97	# 11	59	# 16	56
# 2	87	# 7	86	# 12	68	# 17	67
# 3	76	# 8	75	# 13	75	# 18	78
# 4	65	# 9	64	# 14	86	# 19	89
# 5	54	# 10	53	# 15	95	# 20	96

Review 1 B

# 1	79	# 6	78	# 11	59	# 16	56
# 2	86	# 7	85	# 12	69	# 17	67
# 3	76	# 8	75	# 13	75	# 18	78
# 4	67	# 9	64	# 14	87	# 19	89
# 5	56	# 10	53	# 15	97	# 20	89

Grade 4 Answers Chapter 3

Lesson 13 Review 2 A

# 1	96	# 6	95	# 11	57	# 16	54
# 2	85	# 7	84	# 12	66	# 17	65
# 3	74	# 8	73	# 13	72	# 18	76
# 4	63	# 9	62	# 14	84	# 19	87
# 5	52	# 10	51	# 15	93	# 20	94

Review 2 B

# 1	77	# 6	76	# 11	57	# 16	54
# 2	84	# 7	83	# 12	67	# 17	65
# 3	74	# 8	73	# 13	73	# 18	76
# 4	65	# 9	62	# 14	85	# 19	87
# 5	54	# 10	51	# 15	95	# 20	86

Lesson 14

	a	b	c		a	b	c
# 1	4	12	24	# 21	2	3	4
# 2	8	24	48	# 22	4	6	8
# 3	12	36	72	# 23	6	9	12
# 4	16	48	96	# 24	8	12	16
# 5	20	60	120	# 25	10	15	20
# 6	24	72	144	# 26	12	18	24
# 7	28	84	168	# 27	14	21	28
# 8	32	96	192	# 28	16	24	32
# 9	36	108	216	# 29	18	27	36
# 10	40	120	240	# 30	20	30	40

	a	b	c		a	b	c
# 11	15	30	49	# 31	5	6	7
# 12	30	60	98	# 32	10	12	14
# 13	45	90	147	# 33	15	18	21
# 14	60	120	196	# 34	20	24	28
# 15	75	150	245	# 35	25	30	35
# 16	90	180	294	# 36	30	36	42
# 17	105	210	343	# 37	35	42	49
# 18	120	240	392	# 38	40	48	56
# 19	135	270	441	# 39	45	54	63
# 20	150	300	490	# 40	50	60	70

Grade 4 Answers Chapter 3

Lesson 15

	a	b	c
# 1	2	3	4
# 2	4	6	8
# 3	6	9	12
# 4	8	12	16
# 5	10	15	20
# 6	12	18	24
# 7	14	21	28
# 8	16	24	32
# 9	18	27	36
# 10	20	30	40

	a	b	c
# 21	1	1	1
# 22	2	2	2
# 23	3	3	3
# 24	4	4	4
# 25	5	5	5
# 26	6	6	6
# 27	7	7	7
# 28	8	8	8
# 29	9	9	9
# 30	10	10	10

Lesson 15

	a	b	c
# 11	5	6	7
# 12	10	12	14
# 13	15	18	21
# 14	20	24	28
# 15	25	30	35
# 16	30	36	42
# 17	35	42	49
# 18	40	48	56
# 19	45	54	63
# 20	50	60	70

	a	b	c
# 31	1	1	1
# 32	2	2	2
# 33	3	3	3
# 34	4	4	4
# 35	5	5	5
# 36	6	6	6
# 37	7	7	7
# 38	8	8	8
# 39	9	9	9
# 40	10	10	10

Lesson 16

	a	b	c	d
# 1	1	1	1	1
# 2	3	3	3	3
# 3	5	5	5	5
# 4	7	7	7	7
# 5	9	9	9	9
# 6	11	11	11	11
# 7	13	13	13	13
# 8	15	15	15	15
# 9	17	17	17	17
# 10	19	19	19	19

	a	b	c	d
# 11	1	1	1	1
# 12	3	3	3	3
# 13	5	5	5	5
# 14	7	7	7	7
# 15	9	9	9	9
# 16	11	11	11	11
# 17	13	13	13	13
# 18	15	15	15	15
# 19	17	17	17	17
# 20	19	19	19	19

Grade 4 Answers Chapter 3

Lesson 16

	a	b	c	d		a	b	c	d
# 21	10	6	3	1	# 31	17	20	23	26
# 22	11	7	4	2	# 32	22	26	30	34
# 23	12	8	5	3	# 33	27	32	33	42
# 24	13	9	6	4	# 34	32	38	44	50
# 25	14	10	7	5	# 35	37	44	51	58
# 26	15	11	8	6	# 36	42	50	58	66
# 27	16	12	9	7	# 37	47	56	65	74
# 28	17	13	10	8	# 38	52	62	72	82
# 29	18	14	11	9	# 39	57	68	79	90
# 30	19	15	12	10	# 40	62	74	86	98

Lesson 17 Review 4 A

# 1	126	# 6	712	# 11	162	# 16	711
# 2	27	# 7	100	# 12	72	# 17	102
# 3	24	# 8	94	# 13	45	# 18	99
# 4	13	# 9	777	# 14	23	# 19	480
# 5	149	# 10	810	# 15	146	# 20	800

Review 4 B

# 1	180	# 6	700	# 11	180	# 16	500
# 2	24	# 7	60	# 12	35	# 17	100
# 3	21	# 8	100	# 13	12	# 18	40
# 4	4	# 9	300	# 14	54	# 19	300
# 5	160	# 10	300	# 15	150	# 20	400

Lesson 18 Test 4 A

# 1	162	# 6	714	# 11	216	# 16	713
# 2	18	# 7	95	# 12	24	# 17	96
# 3	30	# 8	99	# 13	28	# 18	94
# 4	13	# 9	703	# 14	23	# 19	465
# 5	181	# 10	800	# 15	174	# 20	800

Test 4 B

# 1	252	# 6	715	# 11	270	# 16	714
# 2	28	# 7	97	# 12	30	# 17	98
# 3	24	# 8	114	# 13	18	# 18	94
# 4	31	# 9	250	# 14	47	# 19	330
# 5	176	# 10	750	# 15	179	# 20	850

Grade 4

Chapter 4 Finding unknowns (100 – 999)

Grade 4 Chapter 4 Contents
Finding unknowns
Lesson 1 - 12
Advanced Problems

Work with lesson 1, read rules and examples, 2) Solve the problems from contents, 3) Check answers in the end of chapter 1, 4) Work with the next lessons 2, 3, 4, 5, …, 5) Solve Test 1 A, 6) Grading Scale: A =10 points, B=9points, C=8 points, D=7 points, F=0-6 points, 7) If you solve 10 problems correctly, you get an A, 8) If have an A, you can move on to next chapter, 9) If your grade is lower than an A, work with the mistakes, then solve Test 1B, 10) When you finish all of the chapters, solve Final Exam in last chapter.

Grade 4 Chapter 4 Lesson 1
Finding unknowns 1

Addition: Addends smaller than sum. Sum greater than any addends.

First number in addition called the addend 2

Second number in addition called the addend 3

Result of addition called the sum 5

Addend + Addend = Sum $2 + 3 = 5$ $5 > 2$ $5 > 3$

First Addend=Sum–Second Addend. Second Addend=Sum–First Addend

$? + 3 = 5$ $? = 5 - 3$ $? = 2$ $2 + ? = 5$ $? = 5 - 2$ $? = 3$

Subtraction: Subtrahend and difference less than minuend.

Minuend greater than subtrahend or difference. $5 > 2$ $5 > 3$

First number in subtraction called the minuend 5

Second number in subtraction called the subtrahend 2

Result of subtraction called the difference 3

Minuend – Subtrahend = Difference $5 - 2 = 3$

Minuend = Subtrahend + Difference. Subtrahend = Minuend – Difference

$? - 2 = 3$ $? = 3 + 2$ $? = 5$ $5 - ? = 3$ $? = 5 - 3$ $? = 2$

Review 1 A. Put correct number instead question mark

# 1	14 + ? = 20	# 6	19 + ? = 23	# 11	20 - ? = 6	# 16	23 - ? = 15
# 2	? + 4 = 21	# 7	? + 4 = 22	# 12	? – 6 = 15	# 17	? – 5 = 17
# 3	19 + ? = 24	# 8	24 + ? = 29	# 13	24 - ? = 16	# 18	29 - ? = 13
# 4	17 + ? = 25	# 9	19 + ? = 28	# 14	? – 6 = 19	# 19	? – 10 = 28
# 5	? + 8 = 26	# 10	18 + ? = 27	# 15	? – 9 = 17	# 2 0	27 - ? = 9

Review 1 B. Complete the tables

1 ? + 15 = 50 ? = 50 – 15 ? = 35 50 > 15 50 > 35

	# 1	# 2	# 3	# 4	# 5	# 6	# 7	# 8	# 9	# 10
Addend	?	16	17		19		16	27		19
Addend	15	?	37	18		26		28	29	
Sum	50	52		56	58	51	53		57	59

11 ? – 39 = 19 ? = 39 + 19 ? = 58 58 > 39 58 > 19

	#11	#12	#13	#14	#15	#16	#17	#18	#19	#20
Minuend	?	56	54		50		57	55		51
Subtrahend	39	?	17	36		42		28	27	
Difference	19	38		16	33	17	28		26	29

Grade 4 Chapter 4 Lesson 1
Finding unknowns 1

a) **Addition:** $2 + 3 = 5$ | $2 < 5$ | $3 < 5$ | $5 > 2$ | $5 > 3$

$? + 3 = 5$ | $? = 5 - 3$ | $? = 2$ | $2 + ? = 5$ | $? = 5 - 2$ | $? = 3$

b) **Subtraction:** $5 – 2 = 3$ | $2 < 5$ | $3 < 5$ | $5 > 2$ | $5 > 3$

$? – 2 = 3$ | $? = 3 + 2$ | $? = 5$ | $5 - ? = 3$ | $? = 5 - 3$ | $? = 2$

Review 1 C
Put correct number instead question mark

# 1	$? + 39 = 60$	# 6	$23 + ? = 61$	# 11	$61 - ? = 28$	# 16	$? - 31 = 29$
# 2	$? + 37 = 62$	# 7	$27 + ? = 63$	# 12	$63 - ? = 26$	# 17	$? - 35 = 37$
# 3	$? + 35 = 64$	# 8	$26 + ? = 65$	# 13	$65 - ? = 39$	# 18	$? – 29 = 35$
# 4	$? + 29 = 66$	# 9	$49 + ? = 67$	# 14	$67 - ? = 39$	# 19	$? - 37 = 29$
# 5	$? + 29 = 68$	# 10	$29 + ? = 69$	# 15	$69 - ? = 49$	# 20	$? – 39 = 29$

Work with examples

a) $? + 59 = 85$ b) $37 + ? = 86$

	a	b	c	d	e
Addend	?	37	48	59	60
Addend	59	?	39	29	29
Sum	85	86	87	88	89

f) $? – 26 = 59$ g) $86 - ? = 37$

	f	g	h	i	j
Minuend	?	86	87	88	89
Subtrahend	26	?	48	29	60
Difference	59	37	39	59	29

Review 1 D Complete the tables

	#1	#2	#3	#4	#5	#6	#7	#8	#9	#10
Addend		19	23		29	59		67	79	
Addend	73		66	75		33	49		20	57
Sum	91	93		97	90		94	96		98

	#11	#12	#13	#14	#15	#16	#17	#18	#19	#20
Minuend		92	91		94	96		97	98	
Subtrahend	19		29	59		57	29		40	65
Difference	71	33		34	49		66	49		34

Grade 4 Chapter 4 Lesson 2
Finding unknowns 2

First number in addition called the addend		**2**		
Second number in addition called the addend		**3**		
Result of addition called the sum		**5**		
Addend + Addend = Sum		**2 + 3 = 5**		
First Addend = Sum – Second Addend	**? + 3 = 5**	**? = 5 – 3**	**? = 2**	
Second Addend = Sum – First Addend	**2 + ? = 5**	**? = 5 – 2**	**? = 3**	

Review 1 A. Put right number instead question mark

	a	b	c
# 1	? + 124 = 420	297 + ? = 423	298 + 128 = ?
# 2	? + 196 = 433	236 + ? = 431	235 + 199 = ?
# 3	? + 145 = 442	296 + ? = 443	295 + 149 = ?
# 4	? + 195 = 452	256 + ? = 454	255 + 196 = ?
# 5	? + 166 = 460	299 + ? = 463	296 + 165 = ?
# 6	? + 195 = 472	276 + ? = 474	275 + 196 = ?
# 7	? + 186 = 480	299+ ? = 483	297 + 185 = ?
# 8	? + 195 = 492	296+ ? = 490	295 + 196 = ?
# 9	? + 129 = 428	234+ ? = 432	297 + 148 = ?
# 10	? + 167 = 462	279+ ? = 473	294 + 199 = ?

First number in subtraction called minuend		**5**		
Second number in subtraction called subtrahend		**2**		
Result of subtraction called difference		**3**		
Minuend – Subtrahend = Difference		**5 – 2 = 3**		
Minuend = Subtrahend + Difference	**? – 2 = 3**	**? = 3 + 2**	**? = 5**	
Subtrahend = Minuend – Difference	**5 - ? = 3**	**? = 5 – 3**	**? = 2**	

Put right number instead question mark

	a	b	c
# 11	? – 194 =226	423 – ? = 227	426 –198 = ?
# 12	? – 196=237	431– ? = 236	434– 199 = ?
# 13	? – 197=245	443– ? = 247	444–195 = ?
# 14	? – 195=257	454– ? = 256	451–196 = ?
# 15	? - 194=266	463- ? = 264	464- 196 = ?
# 16	? – 195=277	474 –? = 276	471–196 = ?
# 17	? – 194=286	483– ? = 284	482–197 = ?
# 18	? – 195=297	490– ? = 296	491–196 = ?
# 19	? – 199=229	432– ? = 234	445–197 = ?
# 20	? - 195=267	473- ? = 279	493- 199 = ?

Grade 4 Chapter 4 Lesson 2
Finding unknowns 2

First number in addition called the addend		2		
Second number in addition called the addend		3		
Result of addition called the sum		5		
Addend + Addend = Sum		2 + 3 = 5		
First Addend = Sum – Second Addend	? + 3 = 5	? = 5 – 3	? = 2	
Second Addend = Sum – First Addend	2 + ? = 5	? = 5 – 2	? = 3	

Review 1 B. Put right number instead question mark

	a	b	c
# 1	? + 124 = 520	397 + ? = 523	398 + 128 = ?
# 2	? + 196 = 533	336 + ? = 531	335 + 199 = ?
# 3	? + 145 = 542	396 + ? = 543	395 + 149 = ?
# 4	? + 195 = 552	356 + ? = 554	355 + 196 = ?
# 5	? + 166 = 560	399 + ? = 563	396 + 165 = ?
# 6	? + 195 = 572	376 + ? = 574	375 + 196 = ?
# 7	? + 186 = 580	399+ ? = 583	397 + 185 = ?
# 8	? + 195 = 592	396+ ? = 590	395 + 196 = ?
# 9	? + 129 = 528	334+ ? = 532	397 + 148 = ?
# 10	? + 167 = 562	379+ ? = 573	394 + 199 = ?

First number in subtraction called minuend		5		
Second number in subtraction called subtrahend		2		
Result of subtraction called difference		3		
Minuend – Subtrahend = Difference		5 – 2 = 3		
Minuend = Subtrahend + Difference	? – 2 = 3	? = 3 + 2	? = 5	
Subtrahend = Minuend – Difference	5 - ? = 3	? = 5 – 3	? = 2	

Put right number instead question mark

	a	b	c
# 11	? – 194 =326	523 – ? = 327	526 –198 = ?
# 12	? – 196=337	531– ? = 336	534 – 199 = ?
# 13	? – 197=345	543– ? = 347	544 –195 = ?
# 14	? – 195=357	554– ? = 356	551 –196 = ?
# 15	? - 194=366	563- ? = 364	564 -196 = ?
# 16	? – 195=377	574 –? = 376	571 –196 = ?
# 17	? – 194=386	583– ? = 384	582 –197 = ?
# 18	? – 195=397	590– ? = 396	591 –196 = ?
# 19	? – 199=329	532– ? =334	545 –197 = ?
# 20	? - 195=367	573- ? = 379	593 – 199 = ?

Grade 4 Chapter 4 Lesson 3
Finding unknowns 3

First number in addition called the addend	2		
Second number in addition called the addend	3		
Result of addition called the sum	5		
Addend + Addend = Sum	2 + 3= 5		
First Addend = Sum – Second Addend	? + 3 = 5	? = 5 – 3	? = 2
Second Addend = Sum – First Addend	2 + ? = 5	? = 5 – 2	? = 3

Review 1 A. Put right number instead question mark

	a	b	c
# 1	? + 124 = 620	497 + ? = 623	498 + 128 = ?
# 2	? + 196 = 633	436 + ? = 631	435 + 199 = ?
# 3	? + 145 = 642	496 + ? = 643	495 + 149 = ?
# 4	? + 195 = 652	456 + ? = 654	455 + 196 = ?
# 5	? + 166 = 660	499 + ? = 663	496 + 165 = ?
# 6	? + 195 = 672	476 + ? = 674	475 + 196 = ?
# 7	? + 186 = 680	499+ ? = 683	497 + 185 = ?
# 8	? + 195 = 692	496+ ? = 690	495 + 196 = ?
# 9	? + 129 = 628	434+ ? = 632	497 + 148 = ?
# 10	? + 167 = 662	479+ ?=673	494 + 199 = ?

First number in subtraction called minuend	5		
Second number in subtraction called subtrahend	2		
Result of subtraction called difference	3		
Minuend – Subtrahend = Difference	5 – 2 = 3		
Minuend = Subtrahend + Difference	? – 2 = 3	? = 3 + 2	? = 5
Subtrahend = Minuend – Difference	5 - ? = 3	? = 5 – 3	? = 2

Put right number instead question mark

	a	b	c
# 11	? – 194 =426	623 –? = 427	626 – 198 = ?
# 12	? – 196=437	631– ? = 436	634 – 199 = ?
# 13	? – 197=445	643– ? = 447	644 – 195 = ?
# 14	? – 195=457	654– ? = 456	651 – 196 = ?
# 15	? - 194=466	663- ? = 464	664 – 196 = ?
# 16	? – 195=477	674 –? = 476	671 – 196 = ?
# 17	? – 194=486	683– ? = 484	682 – 197 = ?
# 18	? – 195=497	690– ? = 496	691 – 196 = ?
# 19	? – 199=429	632– ? =434	645 – 197 = ?
# 20	? - 195=467	673- ? = 479	693 – 199 = ?

Grade 4 Chapter 4 Lesson 3
Finding unknowns 3

First number in addition called the addend	**2**		
Second number in addition called the addend	**3**		
Result of addition called the sum	**5**		
Addend + Addend = Sum	**2 + 3= 5**		
First Addend = Sum – Second Addend	**? + 3 = 5**	**? = 5 – 3**	**? = 2**
Second Addend = Sum – First Addend	**2 + ? = 5**	**? = 5 – 2**	**? = 3**

Review 1 B. Put right number instead question mark

	a	b	c
# 1	? + 122 = 717	596 + ? = 720	597 + 126 = ?
# 2	? + 197 = 735	537 + ? = 733	536 + 195 = ?
# 3	? + 143 = 741	597 + ? = 742	596 + 147 = ?
# 4	? + 197 = 755	557 + ? = 752	556 + 198 = ?
# 5	? + 168 = 765	594 + ? = 760	599 + 164 = ?
# 6	? + 197 = 775	577 + ? = 772	576 + 198 = ?
# 7	? + 189 = 785	594 + ? = 780	599 + 184 = ?
# 8	? + 197 = 795	597 + ? = 792	596 + 194 = ?

First number in subtraction called minuend	**5**		
Second number in subtraction called subtrahend	**2**		
Result of subtraction called difference	**3**		
Minuend – Subtrahend = Difference	**5 – 2 = 3**		
Minuend = Subtrahend + Difference	**? – 2 = 3**	**? = 3 + 2**	**? = 5**
Subtrahend = Minuend – Difference	**5 - ? = 3**	**? = 5 – 3**	**? = 2**

Put right number instead question mark

	a	b	c
# 9	? – 195 = 522	720 –? = 526	723 – 196 = ?
# 10	? – 197 = 538	733 – ? = 537	731 – 195 = ?
# 11	? – 198 = 543	742 – ? = 545	743 – 196 = ?
# 12	? – 197 = 558	752 – ? = 557	754 – 198 = ?
# 13	? – 197 = 568	760 - ? = 566	763 – 199 = ?
# 14	? – 197 = 578	772 –? = 577	774 – 198 = ?
# 15	? – 196 = 589	780 – ? = 586	783 – 199 = ?
# 16	? – 197 = 598	792– ? = 597	790 – 194 = ?

Grade 4 Chapter 4 Lesson 4
Finding unknowns 4

Rules and Examples

First number in addition called the addend	**2**
Second number in addition called the addend	**3**
Result of addition called the sum	**5**

Addend + Addend = Sum $2 + 3 = 5$

First Addend = Sum – Second Addend

$? + 3 = 5$ $? = 5 - 3$ $? = 2$

Second Addend = Sum – First Addend

$2 + ? = 5$ $? = 5 - 2$ $? = 3$

Complete the tables

 # 1

	a	b	c	d	e
Addend		596	597		599
Addend	122		126	128	
Sum	717	720		726	728

 # 2

Addend		537	536		534
Addend	197		195	199	
Sum	735	733		734	732

 # 3

Addend		597	596		597
Addend	143		147	149	
Sum	741	742		744	745

 # 4

Addend		557	556		554
Addend	197		198	196	
Sum	755	752		751	753

Grade 4 Chapter 4 Lesson 4
Finding unknowns 4

Rules and Examples

First number in addition called the addend **2**
Second number in addition called the addend **3**
Result of addition called the sum **5**

Addend + Addend = Sum **2 + 3 = 5**
First Addend = Sum – Second Addend
? + 3 = 5 ? = 5 – 3 ? = 2
Second Addend = Sum – First Addend
2 + ? = 5 ? = 5 – 2 ? = 3

Complete the tables

 # 5 a b c d e

Addend		594	599		595
Addend	168		164	165	
Sum	765	760		761	762

 # 6 a

a) ? + 197 = 775

Addend	?	577	576		579
Addend	197		198	196	
Sum	775	772		771	773

 # 7 b

b) 594 + ? = 780

Addend		594	599		595
Addend	189	?	184	185	
Sum	785	780		782	781

 # 8

Addend	?	597	596	?	594
Addend	197	?	194	196	?
Sum	795	792	?	791	793

Grade 4 Chapter 4 Lesson 4
Finding unknowns 4

Rules and Examples

First number in subtraction called minuend **5**
Second number in subtraction called subtrahend **2**
Result of subtraction called difference **3**

Minuend – Subtrahend = Difference $5 - 2 = 3$
Minuend = Subtrahend + Difference
$? - 2 = 3$ $? = 3 + 2$ $? = 5$
Subtrahend = Minuend – Difference
$5 - ? = 3$ $? = 5 - 3$ $? = 2$

Complete the tables

 # 9

	a	b	c	d	e
Minuend		720	723		728
Subtrahend	195		196	198	
Difference	522	526		528	529

 # 10

Minuend		733	731		732
Subtrahend	197		195	199	
Difference	538	537		535	534

 # 11

Minuend		742	743		745
Subtrahend	198		196	195	
Difference	543	545		549	548

 # 12

Minuend		752	754		753
Subtrahend	197		198	196	
Difference	558	557		555	554

Grade 4 Chapter 4 Lesson 4
Finding unknowns 4

Rules and Examples

First number in subtraction called minuend **5**
Second number in subtraction called subtrahend **2**
Result of subtraction called difference **3**

Minuend – Subtrahend = Difference **5 – 2 = 3**
Minuend = Subtrahend + Difference
? – 2 = 3 ? = 3 + 2 ? = 5
Subtrahend = Minuend – Difference
5 - ? = 3 ? = 5 – 3 ? = 2

Complete the tables

 # 13

	a	b	c	d	e
Minuend		760	763		762
Subtrahend	197		199	196	
Difference	568	566		568	567

 # 14

	a				
Minuend	?	772	774		773
Subtrahend	197		198	196	
Difference	578	577		575	579

a) ? – 197 = 578

 # 15

		b			
Minuend		780	783		781
Subtrahend	196	?	199	197	
Difference	589	586		585	586

b) 780 - ? = 586

 # 16

Minuend	?	792	790	?	793
Subtrahend	197	?	194	196	?
Difference	598	597	?	595	594

Grade 4 Chapter 4 Lesson 5
Finding unknowns 5

Rules and Examples

First number in addition called the addend	**2**
Second number in addition called the addend	**3**
Result of addition called the sum	**5**

Addend + Addend = Sum $2 + 3 = 5$

First Addend = Sum – Second Addend

$? + 3 = 5$ $? = 5 - 3$ $? = 2$

Second Addend = Sum – First Addend

$2 + ? = 5$ $? = 5 - 2$ $? = 3$

Complete the tables

 # 1

	a	b	c	d	e
Addend		696	697		699
Addend	122		126	128	
Sum	817	820		826	828

 # 2

Addend		637	636		634
Addend	197		195	199	
Sum	835	833		834	832

 # 3

Addend		697	696		697
Addend	143		147	149	
Sum	841	842		844	845

 # 4

Addend		657	656		654
Addend	197		198	196	
Sum	855	852		851	853

Grade 4 Chapter 4 Lesson 5
Finding unknowns 5

Rules and Examples

First number in addition called the addend	**2**
Second number in addition called the addend	**3**
Result of addition called the sum	**5**

Addend + Addend = Sum $\qquad$ **2 + 3 = 5**

First Addend = Sum – Second Addend

? + 3 = 5 $\qquad$? = 5 – 3 $\qquad$? = 2

Second Addend = Sum – First Addend

2 + ? = 5 $\qquad$? = 5 – 2 $\qquad$? = 3

Complete the tables

 # 5

	a	b	c	d	e
Addend		694	699		695
Addend	168		164	165	
Sum	865	860		861	862

 # 6

				d	
Addend		677	676	?	679
Addend	197		198	196	
Sum	875	872		871	873

d) ? + 196 = 871

 # 7

					e
Addend		694	699		695
Addend	189		184	185	?
Sum	885	880		882	881

e) 695 + ? = 881

 # 8

Addend	?	697	696	?	695
Addend	197	?	194	196	?
Sum	895	892	?	891	894

Grade 4 Chapter 4 Lesson 5
Finding unknowns 5

Rules and Examples

First number in subtraction called minuend	5
Second number in subtraction called subtrahend	2
Result of subtraction called difference	3

Minuend – Subtrahend = Difference   5 – 2 = 3

Minuend = Subtrahend + Difference

? – 2 = 3   ? = 3 + 2   ? = 5

Subtrahend = Minuend – Difference

5 - ? = 3   ? = 5 – 3   ? = 2

Complete the tables

9

	a	b	c	d	e
Minuend		820	823		828
Subtrahend	195		196	198	
Difference	622	626		628	629

10

Minuend		833	831		832
Subtrahend	197		195	199	
Difference	638	637		635	634

11

Minuend		842	843		845
Subtrahend	198		196	195	
Difference	643	645		649	648

12

Minuend		852	854		853
Subtrahend	197		198	196	
Difference	658	657		655	654

Grade 4 Chapter 4 Lesson 5
Finding unknowns 5

Rules and Examples

First number in subtraction called minuend **5**
Second number in subtraction called subtrahend **2**
Result of subtraction called difference **3**

Minuend – Subtrahend = Difference **5 – 2 = 3**
Minuend = Subtrahend + Difference
? – 2 = 3 ? = 3 + 2 ? = 5
Subtrahend = Minuend – Difference
5 - ? = 3 ? = 5 – 3 ? = 2

Complete the tables

 # 13

	a	b	c	d	e
Minuend		860	863		862
Subtrahend	197		199	196	
Difference	668	666		668	667

 # 14

				d	
Minuend		872	874	?	873
Subtrahend	197		198	196	
Difference	678	677		675	679

d) ? + 196 = 675

 # 15

					e
Minuend		880	883		881
Subtrahend	196		199	197	?
Difference	689	686		685	686

e) 881 - ? = 686

 # 16

Minuend	?	892	890	?	893
Subtrahend	197	?	194	196	?
Difference	698	697	?	695	694

Grade 4 Chapter 4 Lesson 6
Finding unknowns 6

Rules and Examples

First number in addition called the addend	**2**
Second number in addition called the addend	**3**
Result of addition called the sum	**5**

Addend + Addend = Sum **2 + 3 = 5**

First Addend = Sum – Second Addend

? + 3 = 5 **? = 5 – 3** **? = 2**

Second Addend = Sum – First Addend

2 + ? = 5 **? = 5 – 2** **? = 3**

Complete the tables

 # 1

	a	b	c	d	e
Addend		796	797		799
Addend	122		126	128	
Sum	917	920		926	928

 # 2

Addend		737	736		734
Addend	197		195	199	
Sum	935	933		934	932

 # 3

Addend		797	796		797
Addend	143		147	149	
Sum	941	942		944	945

 # 4

Addend		757	756		754
Addend	197		198	196	
Sum	955	952		951	953

Grade 4 Chapter 4 Lesson 6
Finding unknowns 6

Rules and Examples

First number in addition called the addend **2**
Second number in addition called the addend **3**
Result of addition called the sum **5**

Addend + Addend = Sum **2 + 3 = 5**
First Addend = Sum – Second Addend
? + 3 = 5 ? = 5 – 3 ? = 2
Second Addend = Sum – First Addend
2 + ? = 5 ? = 5 – 2 ? = 3

Complete the tables

 # 5

	a	b	c	d	e
Addend		794	799		795
Addend	168		164	165	
Sum	965	960		961	962

 # 6 **a) ? + 197 = 975**

	a				
Addend	?	777	776		779
Addend	197		198	196	
Sum	975	972		971	973

 # 7 **b) 794 + ? = 980**

		b			
Addend		794	799		795
Addend	189	?	184	185	
Sum	985	980		982	981

 # 8

Addend	?	797	796	?	794
Addend	197	?	194	196	?
Sum	995	992	?	991	993

Grade 4 Chapter 4 Lesson 6
Finding unknowns 6

Rules and Examples

First number in subtraction called minuend **5**
Second number in subtraction called subtrahend **2**
Result of subtraction called difference **3**

Minuend – Subtrahend = Difference **5 – 2 = 3**
Minuend = Subtrahend + Difference
? – 2 = 3 **? = 3 + 2** **? = 5**
Subtrahend = Minuend – Difference
5 - ? = 3 **? = 5 – 3** **? = 2**

Complete the tables

 # 9

	a	b	c	d	e
Minuend		920	923		928
Subtrahend	195		196	198	
Difference	722	726		728	729

 # 10

Minuend		933	931		932
Subtrahend	197		195	199	
Difference	738	737		735	734

 # 11

Minuend		942	943		945
Subtrahend	198		196	195	
Difference	743	745		749	748

 # 12

Minuend		952	954		953
Subtrahend	197		198	196	
Difference	758	757		755	754

Grade 4 Chapter 4 Lesson 6
Finding unknowns 6

Rules and Examples

First number in subtraction called minuend **5**
Second number in subtraction called subtrahend **2**
Result of subtraction called difference **3**

Minuend – Subtrahend = Difference **5 – 2 = 3**
Minuend = Subtrahend + Difference
? – 2 = 3 ? = 3 + 2 ? = 5
Subtrahend = Minuend – Difference
5 - ? = 3 ? = 5 – 3 ? = 2

Complete the tables

 # 13

	a	b	c	d	e
Minuend		960	963		962
Subtrahend	197		199	196	
Difference	768	766		768	767

 # 14

				d	
Minuend		972	974	?	973
Subtrahend	197		198	196	
Difference	778	777		775	779

d) ? – 196 = 775

 # 15

					e
Minuend		980	983		981
Subtrahend	196		199	197	?
Difference	789	786		785	786

e) 981 - ? = 786

 # 16

Minuend	?	992	990	?	993
Subtrahend	197	?	194	196	?
Difference	798	797	?	795	794

Grade 4 Chapter 4 Lesson 7

Test 4 A

Add numbers

\# 1 $\begin{array}{r} 375 \\ +\ 586 \\ \hline \end{array}$ # 2 $\begin{array}{r} 397 \\ +\ 486 \\ \hline \end{array}$

Put right number instead question mark

\# 3 ? + 459 = 924   # 4 495 + ? = 842

\# 5 525 + 369 = ?

Complete the table

	# 6	# 7	# 8	# 9	# 10
Addend		256	272		329
Addend	497		544	548	
Sum	721	832		836	921

Subtract numbers

\# 11 $\begin{array}{r} 992 \\ -\ 155 \\ \hline \end{array}$ # 12 $\begin{array}{r} 961 \\ -\ 248 \\ \hline \end{array}$

Put right number instead question mark

\# 13 ? – 386 = 513   # 14 868 – ? = 589

\# 15 737 – 248 = ?

Complete the table

	# 16	# 17	# 18	# 19	# 20
Minuend		957	924		858
Subtrahend	528		297	364	
Difference	462	528		527	659

Grade 4 Chapter 4 Lesson 7

Test 4 B

Add numbers

1   289 + 595

2   383 + 569

Put right number instead question mark

3   ? + 597 = 954   # 4   398 +? = 891

5   425 + 559 = ?

Complete the table

	# 6	# 7	# 8	# 9	# 10
Addend		285	325		399
Addend	589		598	478	
Sum	836	842		845	864

Subtract numbers

11   997 - 748

12   972 - 456

Put right number instead question mark

13   ? – 479 = 446   # 14   888 - ? = 629

15   851 - 296 = ?

Complete the tables

	# 16	# 17	# 18	# 19	# 20
Minuend		936	897		819
Subtrahend	546		498	589	
Difference	429	369		269	624

Grade 4 Chapter 4 Lesson 8
Multiplication 1

First number in multiplication called the factor	**2**
Second number in multiplication called the factor	**3**
Result of multiplication called the product	**6**
Factor × Factor = Product	**$2 \times 3 = 6$**

First Factor = Product ÷ Second Factor

$? \times 3 = 6$ $? = 6 \div 3$ $? = 2$ $6 > 2$

Second Factor = Product ÷ First Factor

$2 \times ? = 6$ $? = 6 \div 2$ $? = 3$ $6 > 3$

Put right number instead question mark

	a	b	c	d
# 1	$? \times 1 = 2$	$? \times 1 = 3$	$? \times 1 = 4$	$? \times 1 = 5$
# 2	$2 \times ? = 4$	$3 \times ? = 6$	$4 \times ? = 8$	$5 \times ? = 10$
# 3	$2 \times 3 = ?$	$3 \times 3 = ?$	$4 \times 3 = ?$	$5 \times 3 = ?$
# 4	$? \times 4 = 8$	$? \times 4 = 12$	$? \times 4 = 16$	$? \times 4 = 20$
# 5	$2 \times ? = 10$	$3 \times ? = 15$	$4 \times ? = 20$	$5 \times ? = 25$
# 6	$2 \times 6 = ?$	$3 \times 6 = ?$	$4 \times 6 = ?$	$5 \times 6 = ?$
# 7	$? \times 7 = 14$	$? \times 7 = 21$	$? \times 7 = 28$	$? \times 7 = 35$
# 8	$2 \times ? = 16$	$3 \times ? = 24$	$4 \times ? = 32$	$5 \times ? = 40$
# 9	$2 \times 9 = ?$	$3 \times 9 = ?$	$4 \times 9 = ?$	$5 \times 9 = ?$
# 10	$? \times 10 = 20$	$? \times 10 = 30$	$? \times 10 = 40$	$? \times 10 = 50$

# 11	$? \times 1 = 6$	$? \times 1 = 7$	$? \times 1 = 8$	$? \times 1 = 9$
# 12	$6 \times ? = 12$	$7 \times ? = 14$	$8 \times ? = 16$	$9 \times ? = 18$
# 13	$6 \times 3 = ?$	$7 \times 3 = ?$	$8 \times 3 = ?$	$9 \times 3 = ?$
# 14	$? \times 4 = 24$	$? \times 4 = 28$	$? \times 4 = 32$	$? \times 4 = 36$
# 15	$6 \times ? = 30$	$7 \times ? = 35$	$8 \times ? = 40$	$9 \times ? = 45$
# 16	$6 \times 6 = ?$	$7 \times 6 = ?$	$8 \times 6 = ?$	$9 \times 6 = ?$
# 17	$? \times 7 = 42$	$? \times 7 = 49$	$? \times 7 = 56$	$? \times 7 = 63$
# 18	$6 \times ? = 48$	$7 \times ? = 56$	$8 \times ? = 64$	$9 \times ? = 72$
# 19	$6 \times 9 = ?$	$7 \times 9 = ?$	$8 \times 9 = ?$	$9 \times 9 = ?$
# 20	$? \times 10 = 60$	$? \times 10 = 70$	$? \times 10 = 80$	$? \times 10 = 90$

Grade 4 Chapter 4 Lesson 9
Division 1

Rules and Examples	**$6 \div 2 = 3$**	**$2 \times 3 = 6$**		
First number in division called the dividend	**6**			
Second number in division called the divisor	**2**			
Result of division called the quotient	**3**			
Dividend ÷ Divisor = Quotient	**$6 \div 2 = 3$**	**$6 > 2$**	**$6 > 3$**	
Dividend = Quotient × Divisor	**$? \div 2 = 3$**	**$? = 3 \times 2$**	**$? = 6$**	
Divisor = Dividend ÷ Quotient	**$6 \div ? = 3$**	**$? = 6 \div 3$**	**$? = 2$**	

Put right number instead question mark

	a	b	c	d
# 1	? ÷ 2 = 1	? ÷ 3 = 1	? ÷ 4 = 1	? ÷ 5 = 1
# 2	4 ÷ ? = 2	6 ÷ ? = 2	8 ÷ ? = 2	10 ÷ ? = 2
# 3	6 ÷ 2 = ?	9 ÷ 3 = ?	12 ÷ 4 = ?	15 ÷ 5 = ?
# 4	? ÷ 2 = 4	? ÷ 3 = 4	? ÷ 4 = 4	? ÷ 5 = 4
# 5	10 ÷ ? = 5	15 ÷ ? = 5	20 ÷ ? = 5	25 ÷ ? = 5
# 6	12 ÷ 2 = ?	18 ÷ 3 = ?	24 ÷ 4 = ?	30 ÷ 5 = ?
# 7	? ÷ 2 = 7	? ÷ 3 = 7	? ÷ 4 = 7	? ÷ 5 = 7
# 8	16 ÷ ? = 8	24 ÷ ? = 8	32 ÷ ? = 8	40 ÷ ? = 8
# 9	18 ÷ 2 = ?	27 ÷ 3 = ?	36 ÷ 4 = ?	45 ÷ 5 = ?
# 10	? ÷ 2 = 10	? ÷ 3 = 10	? ÷ 4 = 10	? ÷ 5 = 10

# 11	? ÷ 6 = 1	? ÷ 7 = 1	? ÷ 8 = 1	? ÷ 9 = 1
# 12	12 ÷ ? = 2	14 ÷ ? = 2	16 ÷ ? = 2	18 ÷ ? = 2
# 13	18 ÷ 6 = ?	21 ÷ 7 = ?	24 ÷ 8 = ?	27 ÷ 9 = ?
# 14	? ÷ 6 = 4	? ÷ 7 = 4	? ÷ 8 = 4	? ÷ 9 = 4
# 15	30 ÷ ? = 5	35 ÷ ? = 5	40 ÷ ? = 5	45 ÷ ? = 5
# 16	36 ÷ 6 = ?	42 ÷ 7 = ?	48 ÷ 8 = ?	54 ÷ 9 = ?
# 17	? ÷ 6 = 7	? ÷ 7 = 7	? ÷ 8 = 7	? ÷ 9 = 7
# 18	48 ÷ ? = 8	56 ÷ ? = 8	64 ÷ ? = 8	72 ÷ ? = 8
# 19	54 ÷ 6 = ?	63 ÷ 7 = ?	72 ÷ 8 = ?	81 ÷ 9 = ?
# 20	? ÷ 6 = 10	? ÷ 7 = 10	? ÷ 8 = 10	? ÷ 9 = 10

Grade 4 Chapter 4 Lesson 10
Multiplication 2

$2 \times 3 = 6$ $6 > 3$ $6 > 2$ **Complete the tables**

$? \times 3 = 6$ $? = 6 \div 3$ $? = 2$ $2 \times ? = 6$ $? = 6 \div 2$ $? = 3$

1

	a	b	c	d	e
Factor		2	50		18
Factor	2		2	2	2
Product	20	60		140	

2

Factor		3	51		19
Factor	3		3	3	3
Product	33	93		213	

3

Factor		4	52		28
Factor	4		4	4	4
Product	48	128		288	

4

Factor		5	53		29
Factor	5		5	5	5
Product	65	165		365	

5

	a	b	c	d	e
Factor		6	54		38
Factor	6		6	6	6
Product	84	204		444	

6

a) $? \times 7 = 105$

	a				
Factor	?	7	56		39
Factor	7		7	7	7
Product	105	245		525	

7

b) $8 \times ? = 288$

		b			
Factor		8	56		48
Factor	8	?	8	8	8
Product	128	288		608	

8

Factor	?	9	57	?	49
Factor	9	?	9	9	9
Product	153	333	?	693	?

Grade 4 Chapter 4 Lesson 10
Multiplication 2

$2 \times 3 = 6$ $6 > 3$ $6 > 2$ **Complete the tables**

$? \times 3 = 6$ $? = 6 \div 3$ $? = 2$ $2 \times ? = 6$ $? = 6 \div 2$ $? = 3$

 # 9

	a	b	c	d	e
Factor		2	60		58
Factor	2		2	2	2
Product	40	80		160	

 # 10

Factor		3	61		59
Factor	3		3	3	3
Product	63	123		243	

 # 11

Factor		4	62		68
Factor	4		4	4	4
Product	88	168		328	

 # 12

Factor		5	63		69
Factor	5		5	5	5
Product	115	215		415	

 # 13

	a	b	c	d	e
Factor		6	64		78
Factor	6		6	6	6
Product	144	264		504	

 # 14

				d	
Factor		7	65	?	79
Factor	7		7	7	7
Product	175	315		595	

d) $? \times 7 = 595$

 # 15

		b			
Factor		8	66		88
Factor	8	?	8	8	8
Product	208	368		688	

b) $8 \times ? = 368$

 # 16

Factor	?	9	67	?	89
Factor	9	?	9	9	9
Product	243	423	?	783	?

Grade 4 Chapter 4 Lesson 11
Division 2

$6 \div 2 = 3$ $6 > 3$ $6 > 2$ **Complete the tables**

$? \div 2 = 3$ $? = 3 \times 2$ $? = 6$ $6 \div ? = 3$ $? = 6 \div 3$ $? = 2$

 # 1   a   b   c   d   e

Dividend		30	50		18
Divisor	2		2	2	2
Quotient	5	2		35	

 # 2

Dividend		93	153		57
Divisor	3		3	3	3
Quotient	11	3		71	

 # 3

Dividend		128	208		112
Divisor	4		4	4	4
Quotient	12	4		72	

 # 4

Dividend		165	265		145
Divisor	5		5	5	5
Quotient	13	5		73	

 # 5   a   b   c   d   e

Dividend		204	324		228
Divisor	6		6	6	6
Quotient	14	6		74	

 # 6   a

a) $? \div 7 = 15$

Dividend	?	245	385		273
Divisor	7		7	7	7
Quotient	15	7		75	

 # 7   b

b) $288 \div ? = 8$

Dividend		288	448		384
Divisor	8	?	8	8	8
Quotient	16	8		76	

 # 8

Dividend	?	333	513	?	441
Divisor	9	?	9	9	9
Quotient	17	9	?	77	?

Grade 4 Chapter 4 Lesson 11
Division 2

6 ÷ 2 = 3 6 > 3 6 > 2 **Complete the tables**

? ÷ 2 = 3 ? = 3 × 2 ? = 6 6 ÷ ? = 3 ? = 6 ÷ 3 ? = 2

9

	a	b	c	d	e
Dividend		80	120		116
Divisor	2		2	2	2
Quotient	20	2		80	

10

Dividend		123	183		177
Divisor	3		3	3	3
Quotient	21	3		81	

11

Dividend		168	248		272
Divisor	4		4	4	4
Quotient	22	4		82	

12

Dividend		215	315		345
Divisor	5		5	5	5
Quotient	23	5		83	

13

	a	b	c	d	e
Dividend		264	384		468
Divisor	6		6	6	6
Quotient	24	6		84	

14 b) 315 ÷ ? = 7

		b			
Dividend		315	455		553
Divisor	7	?	7	7	7
Quotient	25	7		85	

15 d) ? ÷ 8 = 86

				d	
Dividend		368	528	?	704
Divisor	8		8	8	8
Quotient	26	8		86	

16

Dividend	?	423	603	?	801
Divisor	9	9	9	9	9
Quotient	27		?	87	?

Grade 4 Chapter 4 Lesson 12
Summary 3

1)Multiplication numbers: for memory <u>(x, x, any)</u>
To multiply numbers, work in any order
$9 \times 8 \times 7 = 72 \times 7 = 504$ $\quad$ $9 \times 8 \times 7 = 9 \times 56 = 504$

2) Work from left to right: for memory <u>(L → R)</u>

a)Multiplication and Division numbers: (× or ÷)
To multiply and divide numbers work from left to right (L → R)
$9 \times 6 \div 3 = 54 \div 3 = 18$ $\quad$ $4 \times 81 \div 9 = 324 \div 9 = 36$

b)Division and Multiplication numbers: (÷ or ×)
To divide and multiply numbers work from left to right <u>(L → R)</u>
$24 \div 6 \times 2 = 4 \times 2 = 8$ $\quad$ $81 \div 9 \times 4 = 9 \times 4 = 36$

d) Division numbers: (÷, ÷)
To divide numbers, work from left to right <u>(L → R)</u>
$24 \div 4 \div 2 = 6 \div 2 = 3$ $\quad$ $810 \div 9 \div 3 = 90 \div 9 = 10$

3) Do Multiplication or Division before Addition or Subtraction
for memory (× or ÷) before (+ or -)
a)Multiplication and Addition numbers: 1) × 2) +
To multiply and add numbers always first multiply than add
$9 \times 6 + 9 = 54 + 9 = 63$ $\quad$ $9 + 9 \times 9 = 9 + 81 = 90$

b)Multiplication and Subtraction numbers: 1) × 2) −
To multiply and subtract numbers always first multiply than subtract
$9 \times 9 - 9 = 81 - 9 = 72$ $\quad$ $90 - 9 \times 9 = 90 - 81 = 9$

c)Division and Addition numbers: 1) ÷ 2) +
To divide and add numbers always first divide than add
$50 \div 5 + 39 = 10 + 39 = 49$ $\quad$ $88 + 171 \div 9 = 88 + 19 = 107$

d)Division and Subtraction numbers: 1) ÷ 2) −
To divide and subtract numbers always first divide than subtract
$108 \div 6 - 9 = 18 - 9 = 9$ $\quad$ $500 - 126 \div 7 = 500 - 18 = 482$

Grade 4 Chapter 4 Lesson 12
Review 4 A (Basic Problems)
Evaluate

# 1	$8 \times 5 \times 6$
# 2	$9 \times 24 \div 8$
# 3	$56 \div 8 \times 4$
# 4	$640 \div 8 \div 4$
# 5	$109 + 9 \times 9$
# 6	$60 \times 8 - 80$
# 7	$240 - 8 \times 5$
# 8	$720 \div 9 - 20$
# 9	$333 + 444 \div 4$
# 10	$360 - 360 \div 6$

Grade 4 Chapter 4 Lesson 12
Review 4 B (Basic Problems)
Evaluate

1 $6 \times 10 \times 10$

2 $7 \times 24 \div 6$

3 $72 \div 9 \times 4$

4 $480 \div 8 \div 4$

5 $104 + 6 \times 6$

6 $80 \times 9 - 20$

7 $135 - 7 \times 5$

8 $630 \div 9 - 10$

9 $444 + 555 \div 5$

10 $560 - 420 \div 7$

Grade 4 Chapter 4 Lesson 12

Final Exam 4 A (Basic Problems)

Evaluate

# 1	$6 \times 5 \times 8$
# 2	$8 \times 27 \div 9$
# 3	$63 \div 9 \times 4$
# 4	$720 \div 9 \div 4$
# 5	$106 + 8 \times 8$
# 6	$70 \times 9 - 30$
# 7	$145 - 9 \times 5$
# 8	$480 \div 8 - 20$
# 9	$222 + 333 \div 3$
# 10	$240 - 240 \div 6$

Grade 4 Chapter 4 Lesson 12
Final Exam 4 B (Basic Problems)
Evaluate

# 1	$8 \times 10 \times 6$
# 2	$6 \times 45 \div 9$
# 3	$81 \div 9 \times 2$
# 4	$810 \div 9 \div 2$
# 5	$111 + 7 \times 7$
# 6	$90 \times 8 - 20$
# 7	$160 - 8 \times 7$
# 8	$640 \div 8 - 20$
# 9	$210 + 280 \div 7$
# 10	$250 - 400 \div 8$

Grading Scale: A =10 points, B=9points, C=8 points, D=7 points, F=0-6 points

Final Exam 4 B (Basic Problems)

# 1	480	# 6	700
# 2	30	# 7	104
# 3	18	# 8	60
# 4	45	# 9	250
# 5	160	# 10	200

Grade 4 Answers Chapter 4

Lesson 1 Review 1 A

# 1	6	# 11	14
# 2	17	# 12	21
# 3	5	# 13	8
# 4	8	# 14	25
# 5	18	# 15	26
# 6	4	# 16	8
# 7	18	# 17	22
# 8	5	# 18	16
# 9	9	# 19	38
# 10	9	# 20	18

Review 1 B

# 1	35	# 11	58
# 2	36	# 12	18
# 3	54	# 13	37
# 4	38	# 14	52
# 5	39	# 15	17
# 6	25	# 16	59
# 7	37	# 17	29
# 8	55	# 18	27
# 9	28	# 19	53
# 10	40	# 20	22

Lesson 1 Review 1 C

# 1	21	# 11	33
# 2	25	# 12	37
# 3	29	# 13	26
# 4	37	# 14	28
# 5	39	# 15	20
# 6	38	# 16	60
# 7	36	# 17	72
# 8	39	# 18	64
# 9	18	# 19	66
# 10	40	# 20	68

Review 1 D

# 1	18	# 11	90
# 2	74	# 12	59
# 3	89	# 13	62
# 4	22	# 14	93
# 5	61	# 15	45
# 6	92	# 16	39
# 7	45	# 17	95
# 8	29	# 18	48
# 9	99	# 19	58
# 10	41	# 20	99

Lesson 2 Review 1 A

	a b c		a b c
# 1	296, 126, 426	# 11	420, 196, 228
# 2	237, 195, 434	# 12	433, 195, 235
# 3	297, 147, 444	# 13	442, 196, 249
# 4	257, 198, 451	# 14	452, 198, 255
# 5	294, 164, 461	# 15	460, 199, 268
# 6	277, 198, 471	# 16	472, 198, 275
# 7	294, 184, 482	# 17	480, 199, 285
# 8	297, 194, 491	# 18	492, 194, 295
# 9	299, 198, 445	# 19	428, 198, 248
# 10	295, 194, 493	# 20	462, 194, 294

Grade 4 Answers Chapter 4

Lesson 2 Review 1 B

	a b c		a b c
# 1	396, 126, 526	# 11	520, 196, 328
# 2	337, 195, 534	# 12	533, 195, 335
# 3	397, 147, 544	# 13	542, 196, 349
# 4	357, 198, 551	# 14	552, 198, 355
# 5	394, 164, 561	# 15	560, 199, 368
# 6	377, 198, 571	# 16	572, 198, 375
# 7	394, 184, 582	# 17	580, 199, 385
# 8	397, 194, 591	# 18	592, 194, 395
# 9	399, 198, 545	# 19	528, 198, 348
# 10	395, 194, 593	# 20	562, 194, 394

Lesson 3 Review 1 A

	a b c		a b c
# 1	496, 126, 626	# 11	620, 196, 428
# 2	437, 195, 634	# 12	633, 195, 435
# 3	497, 147, 644	# 13	642, 196, 449
# 4	457, 198, 651	# 14	652, 198, 455
# 5	494, 164, 661	# 15	660, 199, 468
# 6	477, 198, 671	# 16	672, 198, 475
# 7	494, 184, 682	# 17	680, 199, 485
# 8	497, 194, 691	# 18	692, 194, 495
# 9	499, 198, 645	# 19	628, 198, 448
# 10	495, 194, 693	# 20	662, 194, 494

Lesson 3 Review 1 B

	a b c		a b c
# 1	595, 124, 723	# 9	717, 194, 527
# 2	538, 196, 731	# 10	735, 196, 536
# 3	598, 145, 743	# 11	741, 197, 547
# 4	558, 195, 754	# 12	755, 195, 556
# 5	597, 166, 763	# 13	765, 194, 564
# 6	578, 195, 774	# 14	775, 195, 576
# 7	596, 186, 783	# 15	785, 194, 584
# 8	598, 195, 790	# 16	795, 195, 596

Grade 4 Answers Chapter 4

Lesson 4

	a b c d e		a b c d e
# 1	595, 124, 723, 598, 129	# 9	717, 194, 527, 726, 199
# 2	538, 196, 731, 535, 198	# 10	735, 196, 536, 734, 198
# 3	598, 145, 743, 595, 148	# 11	741, 197, 547, 744, 197
# 4	558, 195, 754, 555, 199	# 12	755, 195, 556, 751, 199
# 5	597, 166, 763, 596, 167	# 13	765, 194, 564, 764, 195
# 6	578, 195, 774, 575, 194	# 14	775, 195, 576, 771, 194
# 7	596, 186, 783, 597, 186	# 15	785, 194, 584, 782, 195
# 8	598, 195, 790, 595, 199	# 16	795, 195, 596, 791, 199

Lesson 5

	a b c d e		a b c d e
# 1	695, 124, 823, 698, 129	# 9	817, 194, 627, 826, 199
# 2	638, 196, 831, 635, 198	# 10	835, 196, 636, 834, 198
# 3	698, 145, 843, 695, 148	# 11	841, 197, 647, 844, 197
# 4	658, 195, 854, 655, 199	# 12	855, 195, 656, 851, 199
# 5	697, 166, 863, 696, 167	# 13	865, 194, 664, 864, 195
# 6	678, 195, 874, 675, 194	# 14	875, 195, 676, 871, 194
# 7	696, 186, 883, 697, 186	# 15	885, 194, 684, 882, 195
# 8	698, 195, 890, 695, 199	# 16	895, 195, 696, 891, 199

Lesson 6

	a b c d e		a b c d e
# 1	795, 124, 923, 798, 129	# 9	917, 194, 727, 926, 199
# 2	738, 196, 931, 735, 198	# 10	935, 196, 736, 934, 198
# 3	798, 145, 943, 795, 148	# 11	941, 197, 747, 944, 197
# 4	758, 195, 954, 755, 199	# 12	955, 195, 756, 951, 199
# 5	797, 166, 963, 796, 167	# 13	965, 194, 764, 964, 195
# 6	778, 195, 974, 775, 194	# 14	975, 195, 776, 971, 194
# 7	796, 186, 983, 797, 186	# 15	985, 194, 784, 982, 195
# 8	798, 195, 990, 795, 199	# 16	995, 195, 796, 991, 199

Lesson 7 Test 4 A

# 1	961	# 6	224	# 11	837	# 16	990
# 2	883	# 7	576	# 12	713	# 17	429
# 3	465	# 8	816	# 13	899	# 18	627
# 4	347	# 9	288	# 14	279	# 19	891
# 5	894	# 10	592	# 15	489	# 20	199

Grade 4 Answers Chapter 4

Lesson 7 Test 4 B

# 1	884	# 6	247	# 11	249	# 16	975
# 2	952	# 7	557	# 12	516	# 17	567
# 3	357	# 8	923	# 13	925	# 18	399
# 4	493	# 9	367	# 14	259	# 19	858
# 5	984	# 10	465	# 15	555	# 20	195

Lesson 8

	a	b	c	d		a	b	c	d
# 1	2,	3,	4,	5	# 11	6,	7,	8,	9
# 2	2,	2,	2,	2	# 12	2,	2,	2,	2
# 3	6,	9,	12,	15	# 13	18,	21,	24,	27
# 4	2,	3,	4,	5	# 14	6,	7,	8,	9
# 5	5,	5,	5,	5	# 15	5,	5,	5,	5
# 6	12,	18,	24,	30	# 16	36,	42,	48,	54
# 7	2,	3,	4,	5	# 17	6,	7,	8,	9
# 8	8,	8,	8,	8	# 18	8,	8,	8,	8
# 9	18,	27,	36,	45	# 19	54,	63,	72,	81
# 10	2,	3,	4,	5	# 20	6,	7,	8,	9

Lesson 9

	a	b	c	d		a	b	c	d
# 1	2,	3,	4,	5	# 11	6,	7,	8,	9
# 2	2,	3,	4,	5	# 12	6,	7,	8,	9
# 3	3,	3,	3,	3	# 13	3,	3,	3,	3
# 4	8,	12,	16,	20	# 14	24,	28,	32,	36
# 5	2,	3,	4,	5	# 15	6,	7,	8,	9
# 6	6,	6,	6,	6	# 16	6,	6,	6,	6
# 7	14,	21,	28,	35	# 17	42,	49,	56,	63
# 8	2,	3,	4,	5	# 18	6,	7,	8,	9
# 9	9,	9,	9,	9	# 19	9,	9,	9,	9
# 10	20,	30,	40,	50	# 20	60,	70,	80,	90

Lesson 10

	a	b	c	d	e		a	b	c	d	e
# 1	10,	30,	100,	70,	36	# 9	20,	40,	120,	80,	116
# 2	11,	31,	153,	71,	57	# 10	21,	41,	183,	81,	177
# 3	12,	32,	208,	72,	112	# 11	22,	42,	248,	82,	272
# 4	13,	33,	265,	73,	145	# 12	23,	43,	315,	83,	345
# 5	14,	34,	324,	74,	228	# 13	24,	44,	384,	84,	468
# 6	15,	35,	392,	75,	273	# 14	25,	45,	455,	85,	553
# 7	16,	36,	448,	76,	384	# 15	26,	46,	528,	86,	704
# 8	17,	37,	513,	77,	441	# 16	27,	47,	603,	87,	801

Grade 4 Answers Chapter 4

Lesson 11

	a	b	c	d	e		a	b	c	d	e
# 1	10,	15,	25,	70,	9	# 9	40,	40,	60,	160,	58
# 2	33,	31,	51,	213,	19	# 10	63,	41,	61,	243,	59
# 3	48,	32,	52,	288,	28	# 11	88,	42,	62,	328,	68
# 4	65,	33,	53,	365,	29	# 12	115,	43,	63,	415,	69
# 5	84,	34,	54,	444,	38	# 13	144,	44,	64,	504,	78
# 6	105,	35,	55,	525,	39	# 14	175,	45,	65,	595,	79
# 7	128,	36,	56,	608,	48	# 15	208,	46,	66,	688,	88
# 8	153,	37,	57,	693,	49	# 16	243,	47,	67,	783,	89

Lesson 12 Review 4 A (Basic Problems)

# 1	240	# 6	400
# 2	27	# 7	200
# 3	28	# 8	60
# 4	20	# 9	444
# 5	190	# 10	300

Review 4 B (Basic Problems)

# 1	600	# 6	700
# 2	28	# 7	100
# 3	32	# 8	60
# 4	15	# 9	555
# 5	140	# 10	500

Lesson 12 Final Exam 4 A (Basic Problems)

# 1	240	# 6	600
# 2	24	# 7	100
# 3	28	# 8	60
# 4	20	# 9	333
# 5	170	# 10	200

Final Exam 4 B (Basic Problems)

# 1	480	# 6	700
# 2	30	# 7	104
# 3	18	# 8	60
# 4	45	# 9	250
# 5	160	# 10	200

Students!! Attention!!

Adults, help please students 1 – 4 grades to read and understand advice.

When I was in 5th grade I did not like math. I am not always understood math. I was 11 years girl. Next 6th grade I got a very good teacher. She explains very clearly and I understood her very well. I was worked very hard and I was not afraid to ask questions. Final Grade in Algebra in 6th grade I got an A. I said in class that I want to be a teacher of math. But some students said that I am not enough smart to be a math teacher.
I said I will be work hard with math. I did it, I got a Master's degree in math and education, I love my students, I was teaching many years, in High School, part-time college math instructor, full-time university math lecturer.

Some people do a simple job, eat, sleep, go to work. People with a bad brain have not so a good job, not so good salary. If you look at TV how someone's dance, you will be not changing your brain. But if you take a dancing class this helps your body and brain ability. The same if you want to use the calculator, but the calculator not always with you. Much better if you learn how you need to do any operations with numbers without a calculator.
If you read this math book you will know what you need to do. This teaches you how to improve your memory, attention, logic thinking, how to read your math books every day, your brain works better. You can work 50 minutes with math, 10 minutes break maybe dance. Do not afraid to do mistakes, fixed and move to the next step. Don't listen to anyone, choose what do you like more.

Students need to know if you work with the brain every day, your brain works better. You need a good brain in any job, in any class. When you finish High School, you need to get any profession, you need to go to college and get a good job and a better salary. You need to learn any subject, its work for the brain. Read the book, do the exercises yourself, try your best, and then check your answer at the end of the chapter.
Be honest with yourself. You need to believe in yourself. The brain is like muscles. People with good brain lives longer. The professors at my university were over 80 years old, wise, and optimistic. It was a great pleasure to talk to them.

Made in the USA
Monee, IL
05 November 2026

67380396R10079